The Math of Money

The Math of Money

Making Mathematical Sense of Your Personal Finances

Morton D. Davis

Copernicus Books
An Imprint of Springer-Verlag

Published in the United States by Copernicus Books,
an imprint of Springer-Verlag New York, Inc.
A member of BertelsmannSpringer Science+Business Media GmbH

COPERNICUS BOOKS
37 East 7th Street
New York, NY 10003
www.copernicusbooks.com

LIBRARY OF CONGRESS CATALOGING-IN-PUBLICATION DATA
Davis, Morton D.
  The math of money : making mathematical sense of your personal finances / Morton D. Davis.
     p.   cm.
   Includes index.
    ISBN 0-387-95078-8 (alk. paper)
     1. Investments.   2. Finance.   3. Interest.   I. Title.
   HG4521.D167   2001
    332.4'01'51—dc21   2001017398

Manufactured in the United States of America.
Printed on acid-free paper.
9 8 7 6 5 4 3 2 1

ISBN 0-387-95078-8   SPIN 10771302

To Mark Rubinstein

whose course on options was not only educational but stimulating

Contents

Introduction

This book reflects one mathematician's view of certain areas of economics and finance. It is not a how-to book, it is not exhaustive or rigorous, and it comes with no guarantee of instant wealth. It tries to maximize the use of the reader's imagination and minimize rote calculation. Although some mathematical background is assumed, ordinary high school algebra will usually suffice. Whenever possible, a verbal rather than a mathematical explanation is given, but where formulas serve a useful purpose they are introduced without derivation or apology.

We are primarily concerned with concepts rather than applications, so many topics of practical importance are omitted or papered over—we do not worry about taxes, commissions, and fees and we assume an ideal world in which assets can always be bought at the same price for which they can be sold, and where loans have the same interest rate whether you make them or take them.

On the other hand, we sniff out paradoxes and anomalies that challenge our intuition. At the beginning of each chapter we generally pose a few problems that are discussed later in the text. The reader is invited to use his "common sense" to find solutions to these problems. Those who accept the challenge may find their intuition is at odds with the solution more often than they expect. We hope the solutions surprise the readers and whet their appetites for further exploring.

Does this mean that the book is of no practical interest? Not at all. Despite its idealistic assumptions and restrictions, it can be very useful in the real world. Consider a few examples:

A venture capitalist has an opportunity to invest repeatedly in enterprises in which he is very familiar. He knows the probability of success and the financial consequences of both success and failure. He has a fixed amount of money for his purpose and wants to know what fixed percentage of capital he should invest in each enterprise so his capital will grow most rapidly. (This general situation is not uncommon; an analogous problem faces a day trader on the stock market and an insurance company considering what policies to issue.) Specifically, suppose for every dollar he risks, he either gains two more dollars or loses his dollar, and that each contingency occurs half the time, on average.

The two-to-one odds seem very generous and many people would be highly tempted to risk a high percentage of their money on each investment. If they invest a great many times and invest more than half their money every time, however, they are almost certain to be losers in the long run (although their *average* return may be very high). The mathematically sophisticated, confident that favorable odds guarantee they will be winners in the long run, will be disappointed. It turns out that the optimal amount to bet is a modest 25 percent of one's money if the goal is to increase one's capital at the most rapid rate. This is a counterintuitive, and very practical, result.

Another counterintuitive result from the chapter on options: two companies, A and B, have stocks selling for $100. Two call options, one on A and another on B, expire in one year, and have the same strike price of $65. Both A and B are hoping to sign particular contracts and the consequences of these contracts are fairly accurately estimated. If A is successful, its stock will be worth $160 in a year; if not, it will be worth $60. If B is successful, its stock will be worth $120 in a year; if not, it will be worth $40. Which call is the more valuable? Not only is the answer counterintuitive but it turns out that neither the probability of getting the contract nor attitude toward risk has to be considered.

A final example. After dealing with a broker for many years you find that his batting record on the stock market is 90 percent. You are interested in finding a "miracle stock"—one that triples in a year—and you know only one stock in a thousand is a miracle stock. You pick a stock at random and you know that if you ask your broker if it is a miracle stock, he will say "yes" 90 percent of the time if it is one, and he will say "yes" 10 percent of the time if it is not. How likely is it that you really picked a miracle stock at random if your broker confirms that it is one? Would you say 0.9? It is not even a hundredth of that.

In the chapter on interest rates we observe that an ordinary, garden-variety installment payment plan may have an interest rate that is about twice what it purports to be, and that lotteries paid out over extended time periods may actually be worth less than half their nominal amount. We also discuss some of the risks associated with "riskless" bonds as well as the characteristics of bonds that you should look for to minimize that risk. We discuss practical ways of lowering mortgage payments and the problems facing a retiree who has either a lump sum or a pension from which he must support himself after retirement without changing his standard of living.

This book is intended to be for the general reader rather than to be a textbook. With the exception of a few asides to the more mathematically inclined, the level of quantitative sophistication expected of the reader is modest. But, though the reader is assumed to have little knowledge, he/she obviously should have some curiosity about, and interest in, things economic.

Chapter 1

Investment Strategies

Test Your Intuition 1

"A Sure Thing"

Suppose you have the opportunity to make the following bet as often as you like. You choose to bet $\$X$, a coin is tossed and if heads turns up, you lose your $\$X$; if tails, you win $\$2X$. The probability of getting heads is 50/50.

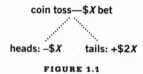

coin toss—$\$X$ bet

heads: $-\$X$ tails: $+\$2X$

FIGURE 1.1

Assume that you start with $100 and make 5,000 bets. To avoid losing all your money, rather than betting a fixed amount you bet a fixed percentage of the money that you have at the time you make your bet, that is, you always bet 75 percent of your capital.

Your first bet would be $75, or 75 percent of $100. If you lose, you would be down $75 and have $25 left; your second bet therefore would be 75 percent of that or $18.75. If you win your first bet, you would win $150 and have a total of $250. Your second bet would be 75 percent of $250 or $187.50.

{1} If you made 5,000 bets using the betting strategy we just described,

(a) What is the average number of dollars you would have when you finished betting?

    i. 10,000   **ii.** 1,000,000,000

    iii. a 1 followed by 690 zeros.

(b) What is the probability that you will have at least $100 when you finish betting—the amount of money that you had when you started?

    i. 0.9999   **ii.** 0.95

    iii. $1/X$ where X is a 1 followed by 44 zeros.

(c) Instead of betting 75 percent of your money all the time, what percentage of your money should you bet to maximize your long-term capital growth?

    i. 50   ii. 25   iii. 10

{2} You start making identical $1 bets repeatedly with $50 in your pocket and resolve to quit when you have either lost all your money or have won enough so that you have a total of $100. The probability of winning any particular bet is 0.45.

(a) What is the probability of leaving a winner?

    i. 0.450   ii. 0.123   iii. 0.00004

(b) How would this answer change if you bet $10 rather than $1 each time?

    i. 0.475   ii. 0.268   iii. 0.150

{3} Suppose the probability of winning each bet in question 2 was 0.49 rather than 0.45.

(a) If you start with $50, always bet $1, and stop when you either have $100 or you have nothing, what is the probability of leaving a winner?

    i. 0.451   ii. 0.351   iii. 0.119

(b) What if you started with $87 instead of $50?

    i. 0.831   ii. 0.732   iii. 0.587

Answers on page 187.

"Would you tell me, please, which way I ought to go from here?" [Alice asked]

"That depends a great deal on where you want to get to," said the Cat.

"I don't much care where_____" said Alice.

"Then it doesn't matter which way you go," said the Cat.

Alice in Wonderland, Lewis Carroll

To invest successfully, folklore has it, you must first define your goals and then fashion an investment strategy to achieve them. But what are we to make of this? Most people have no specific goals other than a vague wish to get as rich as possible, as soon as possible. And in the process they try to emulate the comedian/philosopher Will Rogers by buying a stock and selling it when the price goes up. (And if the price goes down? Then they don't buy it.)

We will say more about the process of setting goals and finding ways to achieve them, but in the end it must be understood that they are *your* goals and so you, like Alice in Wonderland, must first decide where you want to go.

Investments, Bets and Speculations

In U.S. Navy tradition official announcements referred to "officers and their ladies" and to "enlisted men and their wives." Although "ladies" and "wives" may have somewhat different connotations, it was pretty well understood that the two were sisters under the skin. Similarly, the terms investment, bet, and speculation have an analogous relationship. While the term "investment" is often used for longer-term, less volatile, and safer financial transactions we will assume here that all of the three terms have the same meaning and use them interchangeably.

We will generally assume that an investment (bet/speculation) has a certain number of possible outcomes, of which exactly one will occur. Each outcome has a certain probability of occurring, and each outcome has certain financial consequences: a gain or a loss to the investor. For an individual, potential investments might be stocks, bonds, a variety of businesses, savings accounts, insurance, casino gambling, lottery tickets, etc.

It will be further assumed that these probabilities and financial consequences are known to the investor or can be estimated by past history or in some other way. We take as given that in a lifetime an investor will have the opportunity to make a particular investment repeatedly, and we will try to determine whether the investment should be considered at all and, if so, how much capital should be committed.

A company selling earthquake or fire insurance that must decide how much to commit in any one geographic location and a casino setting a limit on the size of bets face essentially the same problem.

The Average Value of an Investment— Favorable, Unfavorable and Fair Bets

If you make one investment repeatedly, the single characteristic of that investment that best indicates whether, and how much, money you will make or lose is its *average* or *expected value*. To find the average value of an investment you first list all possible outcomes (the amount you lose or win) and their probabilities. For each outcome, multiply its probability by the amount you win or lose (a loss is a negative profit) and add up the products corresponding to all outcomes.

Stating this formally, suppose an investment has n possible outcomes and the ith outcome yields a profit/loss of g_i with probability p_i. The average profit is $p_1 g_1 + p_2 g_2 + \cdots + p_n g_n$.

TABLE 1.1			
	Win/Loss	Probability	Win/Loss x Probability
Outcome I	$60,000	1/6	$10,000
Outcome II	$90,000	2/6	$30,000
Outcome III	−$50,000	3/6	−$25,000

An example should make this clearer. Suppose you toss a die and win $60,000 if a 6 turns up (outcome I), win $90,000 if either a 4 or 5 turns up (outcome II), or lose $50,000 if a 1 or 2 or 3 turns up (outcome III). The situation is summarized in Table 1.1.

The average profit is then $10,000 + $30,000 − $25,000 = $15,000. Informally, this means that if you made this wager repeatedly, you would expect to win $15,000, on average, each time you made an investment.

To make this formula plausible imagine you made 6,000 such investments. You might then expect that outcomes I, II and III would occur about 1,000, 2,000, and 3,000 times, respectively, in accordance with their probabilities. For all 6,000 investments you would expect to have a profit of (1,000) ($60,000) + (2,000) ($90,000) + (3,000) (−$25,000). To obtain your profit per investment divide by 6,000, which yields ($\frac{1}{6}$) ($60,000) + ($\frac{2}{6}$) ($90,000) + ($\frac{3}{6}$) (−$25,000), and this is precisely the formula for the expected value.

In a Pick-6 lottery, six winning numbers are chosen from 1 to 49. A player pays a dollar and wins a prize if he has enough winning numbers. He gets $1,000,000 for 6 winning numbers, $10,000 for 5 winning numbers and $100 for 4 winning numbers, and nothing otherwise; the probabilities of getting exactly 6, 5 and 4 winning numbers, respectively, are 0.0000000715, 0.0000184 and 0.000969. All of this is summarized in Table 1.2.

Since you pay $1.00 to play the lottery, and you win $0.352 on average each time you play, it costs you $0.648 on average to play.

If the average profit that you make on an investment is zero, the investment is called a *fair* bet. If the average profit is positive, the odds are with you; if the average profit is negative, the odds are against you.

TABLE 1.2			
(A)	**(B)**	**(C)**	**(B) x (C)**
# of Winning			
Numbers	**Probability**	**Payoff**	**Probability x Payoff**
6	.0000000715	$1,000,000	$0.071
5	.0000184	$ 10,000	$0.184
4	.000969	$ 100	$0.097
			$0.352

Let's start by describing a simple betting system where each individual bet is fair—it is an "unbeatable system," publicized over sixty years ago by *Esquire* magazine, that has seduced more than one casual reader.

"A Sure Thing"

In January 1940 *Esquire* published an article describing a remarkable betting system. It was assumed in the article that a bettor had the opportunity to make the same fair bet repeatedly, and could wager as much as he wished. If the bettor risked a dollar, half the time he would lose it, and half the time he would win another dollar. In fact, the bet itself was of no particular interest; the heart of the system was really resource management—determining how much money to risk on each bet. Nobody knows how many people this article enriched, but it inspired the physicist George Gamow to write a short story.[1]

In Gamow's story, the main character reads this get-rich-quick article and is deeply impressed. Although he is mathematically unsophisticated he has no trouble following the "pure and simple mathematics" on which the betting system is based. After some coin-tossing experiments he decides the system "really works."

---

1. George Gamow, "Maxwell's Demon." In: *Mr. Tompkins in Paperback*, Cambridge University Press, 1967, pp. 95–111.

Most betting systems have a flaw, of course, and Gamow's hero quickly spotted the flaw in this one. Since a national magazine has a vast audience, it was more than likely that a great number of other readers would test this system, and unless he acted quickly there was a grave danger that the casinos would run out of money. He quickly persuaded his wife that their best course was to leave immediately for the casinos and make their fortunes before all those institutions went bankrupt. Before we learn the outcome of this stirring drama, let's take a closer look at *Esquire*'s betting system and see why it is so appealing.

Suppose you toss a fair coin (a coin for which heads and tails are equally likely to turn up) repeatedly and are allowed to bet as much money on each toss as you please. What you win when a head turns up you lose when a tail turns up. According to the *Esquire* article you can almost surely make a profit on a sequence of such games as long as you bet the right amount of money on each toss. The rules of the *Esquire* betting system are listed below, along with an example to show how the system works in practice.

The Betting Rules

(1) Before starting to bet, write down these three numbers: "1, 2, 3."

(2) Start by betting the sum of the numbers on the extreme left and the extreme right. Your first bet would then be 1 + 3 = 4.

(3) Whenever you win, erase the number(s) you bet; whenever you lose, write the number representing the amount of your loss at the right end of the list of numbers. If you won the bet in step 2, your list would contain a single number, 2. If you lost, the list would be "1, 2, 3, 4."

(4) Subsequently, if there is only one number on the list, bet it. If there are more than two numbers on the list, bet the sum of the numbers on the extreme left and the extreme right.

TABLE 1.3

List before Bet	Amount Bet	Result	List after Bet	Amount ahead / behind
1, 2, 3	1 + 3 = 4	Loss	1, 2, 3, 4	–4
1, 2, 3, 4	1 + 4 = 5	Loss	1, 2, 3, 4, 5	–9
1, 2, 3, 4, 5	1 + 5 = 6	Win	2, 3, 4	–3
2, 3, 4	2 + 4 = 6	Loss	2, 3, 4, 6	–9
2, 3, 4, 6	2 + 6 = 8	Win	3, 4	–1
3, 4	3 + 4 = 7	Loss	3, 4, 7	–8
3, 4, 7	3 + 7 = 10	Loss	3, 4, 7, 10	–18
3, 4, 7, 10	3 + 10 = 13	Win	4, 7	–5
4, 7	4 + 7 = 11	Win	blank	+6

(5) Stop when all the numbers—the initial three and the ones you added—have been erased.

One Possible Outcome

Table 1.3 summarizes a series of nine bets. Notice that you lost a majority of your bets (you had five losses and only four wins), but you still won six units.

Why the Betting System Must Be Successful

There is an argument to the effect that eventually this betting system will *always* prevail even though each individual bet is fair. The argument is based upon these two assertions: *Every number will eventually be deleted.* Whenever you win a bet you always delete two numbers (unless the list only has a single member) but whenever you lose a bet you only add one number. In the long run there are as many wins as losses so the list must eventually disappear. *When all the numbers are deleted, you will be 6 ahead.* Every time you write a number (representing a loss) there is a corresponding erasure (representing the same size win). The exceptions—the initial 1, 2, 3 you started with— yield a profit of 1 + 2 + 3 = 6. So it seems that given only the opportunity to make the same fair bet repeatedly and a clever scheme of money management you can almost guarantee that you will make a profit after a succession of bets. Before reading further decide whether you buy this argument or not.

Why the System *Cannot* Work in Theory

We mentioned earlier that each individual bet was fair, which means it gives you an average profit of zero. If you make a series of fair bets over time you will tend to break even, on average. It makes no difference how you vary the amount that you bet. Whether you raise the stakes on some bets and lower them on others your average profit will remain zero. If you hear of a betting system that applies clever money management to a series of fair bets that yields a profit on average, don't even listen to the details; it is impossible. If you restrict yourself to fair bets and if your criterion for judging a betting system is the average profit, don't bother to do any calculations; the outcome will always be the same—an average overall profit of zero—and there is no way you can be either clever or stupid in the way you handle your money.

Why, Despite Appearances, the System Does Not Work in Practice

If you were not suspicious about the *Esquire* betting strategy, you should have been. Although no individual bet offered favorable odds the betting strategy seemed to guarantee an almost certain profit—a sow's ear, it seems, was converted into a silk purse. Suppose you start with $1,000 in your pocket and actually apply the betting system. Although winning isn't a certainty—you might have a perverse string of losses—you will find that in practice you do almost always win. *And so you should.* You are risking $1,000 to win $6, and it is not surprising that you win most of the time. Your probability of losing is $6/1006$; if you bet 1,006 times you might expect to lose 6 times and win 1,000 times, so you would just break even. Of course, if you only risked $100 your chance of losing would become $6/106$ and if you only start with $10 your chance of losing would be $6/16$. (In actual practice, if you start with too little capital you cannot employ the betting strategy as we described it.)

Some General Observations About Investment Strategies Using Fair Bets

While it is true that you cannot extract a positive average profit from a series of fair bets by cleverly manipulating the size of your bets, you are not completely helpless. There is a quasi-conservation law (shown below) that you might be able to use to your advantage:

$$\frac{\text{Your probability of losing}}{\text{Your probability of winning}} = \frac{\text{The amount you will win}}{\text{The amount you risk}}$$

So even if you are restricted to making fair bets you still have some control. You can increase your chance of winning if you will settle for a smaller reward or, if your heart is set on a large reward, you can become reconciled to a larger chance of losing.

If the amount you hope to win equals the amount you risk you will have an even chance of success. With the *Esquire* betting system, where you risked $1,000 to gain $6, the probability of winning was very high; that is why the betting system seems to work so well.

Suppose you start with $15 and bet $1 on a coin toss using a fair coin. If you win you leave with $1; if you lose you double your bet. If you keep doubling up when you lose and leave as soon as you win, you have enough to finance four bets. You leave a loser after four successive losing tosses—this has a probability of $1/16$. You stand to lose $1 + $2 + $4 + $8 = $15 and to gain $1. The quasi-conservation law equation yields $1/15 = (1/16) / (15/16)$.

If you start with $1 and hope to win $15 more, you can start by betting all your money and doubling up when you win. Now your probability of success—four successive wins—is $1/16$ and failure is $15/16$; and again, the formula is easily verified.

The relationship of the probabilities of losing and winning to the amounts that you can win and lose is fairly obvious qualitatively (although it is only approximate because the underlying investment may not really be fair). If you buy a lottery ticket, the probability of losing is much greater than that of winning so the amount you hope to win is much greater than what you risk. If you buy $1,000,000 worth of car insurance, you pay much less as a premium—

which implies that you are not likely to collect on your policy during the year. And even if you are not an expert on horse racing you must be aware that a horse that pays $100 for every $2 you risk is less likely to win (at least in the opinion of the betting public) than a horse that only pays $3 for the same bet.

The reason the *Esquire* betting system is so deceptive is that you are risking a great deal to win very little. A more critical test of the betting system is this: if you try to double your money before you lose it all, how likely are you to succeed? From the formula you can see immediately that the probability of losing equals the probability of winning, that is, you are as likely to double your money as you are to lose it all *no matter what system you adopt*. If you want to double your money there is no "right" strategy. Nor, for that matter, is there a "wrong" strategy. All strategies are equivalent and all yield the same probability of success—1 chance in 2.

Usually an investor faces a more hostile world than the one we have described. Casinos, for example, must make a profit to stay in business and so they make the odds unfavorable to the bettor. Any betting strategy based on unfavorably biased bets can only yield losses (on average), as well. The more you bet, the more you lose. In real casinos the *Esquire* betting system hasn't a chance in the long run.

Why Bet Against the Odds

Should a rational person ever bet against the odds? People often do and may be required to do so by law. Adverse casino odds are often outweighed by the pleasure of gambling. In many states automobile insurance is compulsory and common sense dictates that insurance companies must have the odds in their favor if they are to stay in business. You can expect, therefore, that the money paid out to cover automobile accidents or fires is less than the amount paid in premiums. When you decide whether an investment is right you are really deciding whether an investment is right for *you*. This means that when you make an investment, you should heed Lewis Carroll's Cheshire Cat and first decide where it is you want to go. If you need

$10,000 for a life-saving operation but you only have $5,000, you would likely gamble to make up the difference whatever the odds.

In practice people often bet against the odds because they tend to be averse to large risks; they willingly buy fire insurance to avoid a potentially devastating loss even if the premiums they pay exceed what they might expect to pay, given the actual risk of a fire. If we assume a house is insured for $200,000 and fires occur once every 200 years the insurance premium may be $2,000. The insurance company would expect to make a profit since in any given year it would receive $2,000,000 from its 1,000 policy holders but would only expect to pay about $1,000,000 in claims. But a policy holder may nevertheless feel that the coverage is worth the premium. And people often accept adverse odds even when there are no large potential losses; millions of people buy lottery tickets to get a remote chance of winning a large prize. This is fortunate for lottery managers because the money received must not only exceed the prize money; it must pay for the lottery expenses and yield a surplus as well. The lottery odds may not be favorable, but along with your lottery ticket you buy a dream. And who is to say what a dream is worth?

How Should You Bet Against the Odds?

Assume you are in a situation in which you make the same adverse bet repeatedly and leave only when you either have won some fixed amount set in advance or have lost all your money. It would be better of course to have favorable odds, but even with unfavorable odds, you still have choices and can go about it intelligently. Consider the situation mentioned earlier, for example, in which you had $5,000 and needed $5,000 more for a life-saving operation. Do you do better to plunge your entire $5,000 on an all-or-nothing bet, or should you nurse your capital and accumulate your money slowly until you have garnered the additional $5,000? It turns out that if you must bet with adverse odds then "t'were well it was done quickly."

Although you may beat the odds in the short run as time goes by, a loss becomes more and more inevitable. Suppose you start to bet with $50 and leave when you either lose it all or win an additional $50. Suppose you can bet what you please, your probability of success is 0.45, and you either win or lose the amount that you bet. If you make a single $50 bet, you will leave a winner 45 percent of the time. If you just make $10 bets you will leave a winner 27 percent of the time.

With $5 bets you win less than 12 percent of the time and with $1 bets you win about once in 25,000 times. Although this last assertion may seem incredible—you can verify it if you believe the formula we give—it is true; your chances of coming out ahead after making a great many small bets, each with a probability of success of 0.45, is akin to winning the lottery.

If you only make $1 bets, your prospects are bleak. If you start with $90 and hope to leave with $100 your probability of success is 0.134. Even if you start with $96 your probability of getting another $4 before losing your money is 0.448, less than 1 out of 2. You need to start with $97 to make success more likely than failure and even then the probability is only 0.548.

There is a formula, too complex to derive here, that tells you W, which stands for the probability that you will leave a winner, given your starting capital, the size of your bet, and the probability of winning a bet.

$$ W = \frac{([1-p]/p)^{R/U} - 1}{([1-p]/p)^{100/U} - 1} $$

where
* R is the amount of money you bring to the table
* U is the size of each bet
* p is the probability that you will win any particular bet
* W is the probability you will leave the table with $100 rather than with nothing.

If $p = 0.49$, $U = 1$ and $R = 90$ then

$$ W = \frac{(0.51/0.49)^{90/1} - 1}{(0.51/0.49)^{100/1} - 1} = 0.664 $$

This means that if you have $90, always bet $1, and the probability of winning a bet is 0.49 you will walk away with $100, 66 percent of the time and walk away with nothing 34 percent of the time.

The Law of Large Numbers

So it comes to this: when you only bet a few times, anything can happen. As you make more and more bets in an environment of adverse odds, these odds are more and more likely to take their toll. If you want to double your money, you do best to plunge in and then get out as soon as possible. By the same token, if the odds are favorable and you want to double your money (and are not in too much of a hurry), the safest course is to make many small bets. The smaller your bet, the greater the number of bets and the more certain your ultimate success. This is a consequence of *The Law of Large Numbers*.

Imagine that a biased coin turns up heads 40 percent of the time and tails 60 percent. Imagine also, that when heads turns up you win $1 and if tails turns up you lose $1. Your average return would then be −$.20 = ($1) (0.4) + (−$1) (0.6). If you bet twice, your average loss is twice that, and if you bet 10,000 times, your average loss is −$2,000 = (10,000) (−$.20). The average total profit on a series of bets is simply the sum of the average profits of each individual bet.

There is an important difference between betting once and betting repeatedly, however—the more often you bet, the closer the ratio of your *actual* profit and your theoretical *average* profit tend to be. When you bet once you either win or lose a dollar and the ratio of your actual profit to your theoretical profit is either $1 / (−$0.20) = −5 or (−$1) / (−$0.20) = +5; in either case the actual and average profits are far apart. But if you make 10,000 $1 bets the amount you lose will be between $1,850 and $2,150 (a deviation of $7\frac{1}{2}$ percent from the expected loss of $2,000) about 99.8 percent of the time. This means that 499 out of 500 times the ratio of your actual to your average profit will be between 0.925 and 1.075. If you bet only once, you have a reasonable chance of making a profit even though the odds are unfavorable; if you have the same odds but bet 10,000 times, you are virtually certain to lose.

This convergence of the theoretical average profit and the actual profit when a bet is made repeatedly is a consequence of a statistical theorem known as the law of large numbers. Since casino owners and insurance companies make a great many bets/investments (and have favorable odds on each of them) it is this law that enables them to sleep nights.

For insurance companies, gambling houses, and banks—organizations that make investment decisions repeatedly—the most significant feature of an investment is its average return. If the average profit of a repeated investment is negative, the law of large numbers guarantees that an investor with finite resources must ultimately come to grief. The law of large numbers, though easy enough to understand, does not seem to discourage legions of gamblers who combat unfavorable odds with an inventory of betting systems and assorted superstitions. If you persistently make bets against the odds, you cannot offset your disadvantage by varying the amount that you bet. Parlaying a series of losing bets into a winning system is like having a business that loses money on each transaction but makes it up in volume.

Betting with the Odds

We have discussed thus far fair and unfavorable bets but they are not the whole story. We will not go so far as to reassure Virginia that there is a Santa Claus, but there is no doubt that certain investments prove profitable over long periods of time. Though the stock market is at times is disappointing, over long periods it has proven to be profitable. If you want to minimize risk, long-term U.S. government bonds offer a return that almost always exceeds inflation. And many private enterprises have to show a profit or they could not exist very long. Granted that there are many profitable enterprises, how do you fashion a strategy for investing in them?

We observed earlier that it may be prudent to take a chance even when the odds are against you. So you should not be too surprised to learn that rational people often decline to bet when the odds are even, or are heavily in their favor. If you have $3,000,000 and are given a chance to risk it all on an even chance of winning $10,000,000, it certainly is not irrational to decline even though the bet has an average value of $3,500,000.

If you have an opportunity to make the same bet repeatedly and the odds of that bet are in your favor, you have the same situation as the gambling casino that was mentioned earlier. Now, you could take the view that slow and steady wins the race. For example, if you toss a biased coin that gives you a 55 percent chance of success, you are most likely to achieve your goal if you make minuscule bets. If you have $50 and want to double it and the probability of winning each bet is 0.55, you can achieve your goal with probability 0.99996 if you only make $1 bets; if you only make $10 bets your chance of doubling your money would only be 0.732. In general, the smaller your bets, the longer you keep betting, and therefore the greater your chance of doubling your money before you go broke.

But there is a downside to making small bets when the odds are in your favor—opportunities are not unlimited. When you find a potentially lucrative investment you want to exploit it. If, when the odds are in your favor, you bet too little, your money's growth will be more certain but will also be much slower. The problem to consider is this: how do you best resolve the tension between making small bets that virtually guarantee your ultimate success and making larger bets that more effectively exploit the favorable odds and allows your capital to grow faster?

Assume you are no longer restricted to making the same size bet repeatedly. What new strategies are available and how do you measure their effectiveness? The seductive betting strategy described in *Esquire* should make you wary about evaluating strategies using only common sense. If you haven't yet lost faith in common sense, see what your intuition tells you about the investment strategy described in the next section.

Winning on Average, Losing in Practice—Another "Sure Thing"

Suppose you have an opportunity to make the following bet repeatedly. (You first decide how much to bet; we will assume that you choose to bet X dollars.) A fair coin (that turns up heads half the time) is tossed. If heads turns up, you lose $\$X$, if tails, you win $\$2X$. The odds are clearly and overwhelmingly in your favor; you can make this bet as long as you like and you may stop whenever you choose.

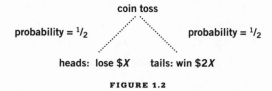

FIGURE 1.2

Suppose you adopt the following strategy: you start with $100 and make 5,000 bets. To avoid losing all your money, rather than betting a fixed amount you bet a fixed percentage of the capital you have when you make the bet. In the present case, let's say you always bet 75 percent of your capital.

To understand the system better, let's look closely at how the betting might go. Your first bet would be $75, 75 percent of $100. If you *lose* the first bet, you will have lost $75 and would have $25 left; your second bet would be 75 percent of that, or $18.75. If you *win* your first bet, you win $150 and you have a total of $250. Your second bet therefore would be 75 percent of $250 or $187.50.

Relying solely on your common sense, and without making any mathematical calculations, try to predict what your situation will be after making 5,000 bets. Specifically:

(a) Would the average amount of money you have when you finish be very large, modest, or very small compared to the $100 with which you started?

(b) Is it almost certain, fairly likely, or virtually impossible that you will have more than the $100 with which you started after you have finished betting?

The odds on each bet are clearly in your favor. For each dollar that you risk, you win two dollars. As you likely guessed, your average profit after 5,000 bets is enormous—in dollars it is more than a 1 followed by 690 zeros.

But this huge average profit is accompanied by a surprising and disconcerting corollary—*the probability that you will have more than the money you started with after 5,000 bets is minuscule*—the probability is less than 1/(100,000,000,000,000,000,000,000,000,000,000,000,000,000,000). This seems paradoxical—how can you make so much money on average and still almost invariably be a loser?

At this point you should reconsider the words of the Cat to Alice. Is it your goal to have a good chance of coming out ahead when the 5,000 bets are over, or to maximize your average profit? The two are not equivalent. If you want to come out ahead, the strategy we described—repeatedly betting 75 percent of your remaining capital—is inappropriate; it attempts to make too much too fast. The odds on each bet are heavily in your favor, but if you push too hard to exploit them by betting too much, you increase your average profit *but at the same time you almost guarantee you will be a loser in the end*. By always betting 75 percent of your capital you can win an enormous amount of money (if you won all 5,000 bets you would have a 1 followed by more than 2,000 zeros in dollars) and your potential losses are only the $100 with which you started. Although your potential gain is much more than your potential loss it comes at a price; you are much more likely to be a loser than a winner. To see this more clearly, take the extreme case in which you bet 100 percent rather than 75 percent of your money. To win anything at all you must win 5,000 bets in a row and the probability of doing that is $1/X$ where X is a 1 followed by 1,505 zeros. But if you *do* win 5,000 bets in a row, you get a one followed by 2,385 zeros in dollars, which makes your average profit very large.

The point—that you should use prudence in exploiting your favorable odds—is important and worth a closer look. Look at what happens if you

TABLE 1.4

Win-Loss Record	Probability	Your Final Capital
0–2	1/4	$ 6.25
1–1	1/2	$ 62.50
2–1	1/4	$ 625.00

use the 75 percent betting system for two bets rather than 5,000 bets. After two bets there are only three possible outcomes, as shown in Table 1.4.

On average, you would leave with capital of $189.06 after two bets, which is almost double the $100 with which you started. (The average is $(^1/_4) (6.25) + (^1/_2) (62.50) + (^1/_4) (625) = \$189.06$). Nevertheless, 75 percent of the time you leave the game a loser. Notice also, that despite the apparently favorable odds, if you win exactly half the time (whether you win the first or second time you bet has no effect on the final outcome), you leave a loser.

Similar statistics for making four bets are shown in Table 1.5. Once again, your average capital when you drop out after four bets is $357.45 which is more than three and a half times your starting capital; but for all that, you leave a loser more than two-thirds of the time. And, again, if you win half of your bets, you leave a loser (and of more than half of your capital at that).

How does a bet with such favorable odds become a losing machine, where you come out behind even though you win half the time? Observe that when you win a bet you multiply your capital by 2.5. (For each $1 that you have, you bet $0.75 and win $1.50, so your $1 becomes $2.50.) When you lose you multiply your capital by 0.25. The effect of a win and a loss in either

TABLE 1.5

Win-Loss Record	Probability	Your Final Capital
0–4	1/16	$ 0.39
1–3	4/16	$ 3.91
2–2	6/16	$ 39.06
3–1	4/16	$ 390.62
4–0	1/16	$3,906.23

order (check it) is the same: it multiplies your capital by (2.5) (0.25) = 0.625. All that matters is the number of wins and losses, not their order. (This is because of the commutative law of multiplication you learned in elementary school: when multiplying numbers to find a product the order of multiplication does not matter.) To come out a winner you must have more wins than losses, and in fact you must win about 60 percent of the time.

If you only bet ten times, you will have four or fewer losses about 38 percent of the time. But if you bet about 400 times, the probability that you will have less than 40 percent losses drops to about 0.00003. The law of large numbers guarantees that in the long run *you are almost certain to be a loser in practice even though you are a winner on average.*

What Is an "Optimal-Sized" Bet?

Having seen examples of deceptively attractive betting strategies, it might be refreshing for you to see how to define (and construct) an optimal one. Once again, imagine a simplified model in which there is one investment with favorable odds that you can make repeatedly. This investment has two possible outcomes: a loss or a win. If you risk $1 and lose, you lose it all; if you win you get $W added to your capital. We assume the probability of winning any single bet is p. Assuming that you are willing to risk $X, the possible outcomes are shown below:

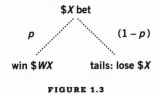

FIGURE 1.3

Say you want to maximize the growth rate of your capital and you decide to always bet the same fraction of your capital. What fraction should that be? Remember, if you bet too much you will leave a loser, while if you bet too little, your capital will grow too slowly.

There is a theory—too complex to describe here—that prescribes what fraction, f, of your capital you should risk on each bet. Given W, the amount that you win for every dollar that you risk, and p, the probability of winning your bet, the fraction of your capital that you should risk is given by the formula[2]

$$f = \frac{PW - 1 + p}{W}$$

In what sense is the fraction f superior to any other fraction, or g? Suppose Paul and Jim start with the same amount of money and make the same bet repeatedly. Paul always bets the fraction f of his capital while Jim always bets the fraction g of his capital. As the number of bets increases, it can be shown that the probability that Paul has more money than Jim gets closer and closer to 1. In other words, eventually you are almost certain to do better using f rather than g if you are patient.

In practice, you generally do not know your potential profits or losses; nor do you know with precision the probabilities involved. But all is not lost. You can often estimate your appropriate f by approximating p and your W by using past history or the nature of the physical world.

In our earlier example, for instance, where $p = 1/2$ and $W = 2$, the optimal f was 0.25. When you bet 75 percent of your capital rather than the recommended 25 percent you had a remote possibility of getting rich fast, but you were much more likely to end up losing money.

The formula for the optimal f is important. To gain insight into its implications it is convenient to write it in a slightly different form: $f = p - (1 - p)/W$. No matter how favorable the odds in your favor (W), f must always be less than p. So if you consider an extreme case in which you

---

2. For the more mathematically inclined, the key is not maximizing your expected winnings but their logarithm. If f is the fraction of your capital that you bet, when you lose a bet ($[1 - p]$ of the time) you multiply your capital by $(1 - f)$, and when you win (p of the time) you multiply your capital by $(1 + fW)$. The average value of the logarithm of your capital after your bet is then $(1 - p)\mathrm{Ln}(1 - f) + p\mathrm{Ln}(1 + fW)$. By elementary calculus this is shown to be maximum when f is given by the formula shown above.

lose $1 with probability 0.9 and you win $100,000,000 with probability 0.1, you should never risk more than a tenth of your capital on a bet. *No matter how large the reward for winning, never risk a greater fraction of your capital than the probability of winning a bet.*

On the other hand, if you have a very high probability of winning, you do well to be more aggressive. So, if the outcome of a bet is that you either lose or win $1 but your probability of winning is 0.99, you should always risk 98 percent of your capital for optimal growth.

Let's take a closer look at the bet we considered earlier, in which $W = 2$ and $p = 0.5$. (This means you win $2 for every $1 you risk, and your chance of winning a bet is the same as losing it.) Observe what happens when ten bets are made with various choices of f. The table entries indicate the amounts you would have after making ten bets assuming that you had $100 initially. Listed on the left are all 11 possible outcomes—from never winning to winning all ten times—and their respective probabilities.

		TABLE 1.6						
$f =$		0.00	0.10	0.25	0.50	0.75	0.90	1.00
$A =$		100	163	325	931	2,416	4,108	5,767
Win/Loss	Probability of Outcome							
0–10	0.001	100	35	6	0	0	0	0
1–9	0.010	100	46	11	0	0	0	0
2–8	0.044	100	62	23	2	0	0	0
3–7	0.117	100	83	45	6	0	0	0
4–6	0.205	100	110	90	25	1	0	0
5–5	0.246	100	147	180	100	10	0	0
6–4	0.205	100	196	360	400	95	5	0
7–3	0.117	100	261	721	1,600	954	135	0
8–2	0.044	100	348	1,442	6,400	9,537	3,778	0
9–1	0.010	100	464	2,883	25,600	95,367	105,785	0
10–0	0.001	100	619	5,767	102,400	953,674	2,961,968	5,904,900

f is the fraction of your capital that you risk on each bet.

A is the average amount of money in dollars that you have left after ten bets.

The value of f reflects how conservative you are. If $f = 0$, you never bet and you leave with the same $100 with which you started. If $f = 1$ you always risk all your capital, you maximize your expected final capital ($5,767), and you get a tiny chance (1 in a 1,024) to win the largest possible prize ($5,904,900). When $f = 1$, however, if you do not win big, you do not win at all. As f increases, your chance of winding up a loser rises but both the size of the big win and your expected value rise as well. Notice that the average profit of $325 using the recommended $f = 0.25$ is considerably smaller than the $5,767 you get with $f = 1$. But with $f = 0.25$ you come out ahead more than 60 percent of the time and with $f = 1$ you win less than 0.1 percent of the time. Ultimately you must decide what you want most at the end of the day—a good chance of a profit or a remote chance of receiving a pot of gold.

Chapter 2

Interest

Test Your Intuition 2

{1} Suppose that Manhattan Island was bought in 1626 for $24 and sold in 1986 for $25,000,000,000,000. The annual rate of interest would be about

    i. 10,000 percent   **ii.** 100 percent   **iii.** 8 percent   **iv.** 2 percent

{2} You borrow $1,000 from a usurious lender at 1 percent per day (compounded daily) and pay nothing until the end of a year; you then pay off the loan and interest in one lump sum. The size of your lump sum payment is about

    i. $3,650   **ii.** $4,650   **iii.** $38,000   **iv.** $105,000

{3} You like to gamble on speculative theatrical productions, where you find you multiply your investment by four 10 percent of the time (a 300 percent profit) and reach breakeven 90 percent of the time. You set aside some speculative money at the beginning of the year and always use all of it (including any profits you have already made) to invest in the next production. What is your average percentage return on each investment (approximately); that is, if you made the same percentage profit on every investment and received the same total profit as you are getting now, what would your profit percentage per investment be?

    i. 5   **ii.** 15   **iii.** 30   **iv.** 60

Answers on page 187.

Interest—The Cost of Renting Money

Interest affects many aspects of our lives in one way or another. When you take out a mortgage, buy on the installment plan, purchase a bond, decide when to start collecting your Social Security payments, or negotiate with your neighborhood usurer, you are involved with interest. Despite these frequent encounters, misconceptions about it abound, and they can be costly.

A dollar that you get today is worth more than a dollar that you get tomorrow, and this is so even if tomorrow's receipt is certain. So when you borrow from a bank you repay not only the amount that you borrowed (the principal), but an additional bonus as well (the interest). The amount of interest that you repay depends upon how much you borrowed, the prevailing interest rate, and the amount of time you are allowed to keep the borrowed money.

Calculating Interest—A Common Error

Suppose you buy a $600 dishwasher and, rather than pay the full $600 at once, you make 13 consecutive monthly payments of $50, with the first payment at the time of purchase. You borrow $600 and repay 13 × $50 = $650, so you pay $50 in interest. Since it takes a year to repay the loan it might seem that your annual interest rate is $50 / $600 or $8^1/_3$ percent. This simple calculation substantially underestimates the real interest rate; in fact, the actual annual interest rate is almost twice as much as $8^1/_3$ percent.

When you lend money you expect to be repaid more than the amount you borrowed, that is, you generally receive interest on your loan. The interest you earn on money is similar to rent on a house—it is a charge for usage; the longer you keep the borrowed money, the more interest you should pay.

If you really borrowed $600 at $8^{1}/_{3}$ percent interest, you would be entitled to keep all of the money for a full year before repaying any part of the principal or interest. In effect, when you buy the dishwasher you borrow the money for about half a year. It is true that you keep your last payment a full year before repaying it but your first payment is made immediately. Very roughly speaking, the first and last payments, taken together, constitute a loan for approximately six months, on average. Similarly, your second and next-to-last payments, taken together, are also repaid after six months, on average. By pairing off the earlier and later payments it should be clear that the real duration of your loan (the average time it took to repay your loan) is more like six months than a year. Effectively, you are paying $8^{1}/_{3}$ percent interest on a six-month loan so you would expect to pay more than twice that if you borrowed the money for a full year.

When you rent a house for six months you don't pay rent for a whole year nor should you pay a year's interest for what is essentially a six-month loan. If you borrow $600 at this rate for a year, you will accrue $600 / 12 = $50 interest in six months and thus owe $650 after six months. Your interest during the second six months would be $650 / 12 = $54.17. Your total interest on the $600 loan would be $50 + $54.17 = $104.17 and the real annual interest rate would be (104.17 / 600) (100) = 17.36 percent.

Simple and Compound Interest—A Distinction With a Big Difference

Suppose that you invest $1,000 in a company that offers to pay you 10 percent interest per year. If you invest your money and let the interest accumulate for 40 years, the amount of money you accumulate may be as little as $5,000 or as much as $54,598. This huge discrepancy arises from one critical variable in how the interest is calculated: the length of time you must wait until the interest that you earn itself earns interest.

To get an idea of the effect that frequency of compounding has on the amount you earn in interest, assume that you invest $1,000 at various rates

Interest Rate	Never	1 Year	3 Months	1 Day	Continuous
			Compounding Period		
2%	$1,800	$ 2,208	$ 2,221	$ 2,225	$ 2,226
6%	$3,400	$10,286	$10,828	$11,021	$11,023
10%	$5,000	$45,259	$51,978	$54,568	$54,598

TABLE 2.1

of interest for 40 years with varying periods of compounding and observe in Table 2.1 how the amount you earn varies.

"Never" compounding is equivalent to simple interest and "continuous" compounding is instantaneous. Notice the diminishing returns as the compounding period gets smaller.

At one extreme is simple interest, where the interest earned never earns interest. If you receive 10 percent simple interest you will have a total of $1,100 at the end of one year, $1,200 at the end of two years, and $1,000 + 100n at the end of n years. If you earn compound interest and interest is compounded annually, that means that during the first year you receive simple interest; after the first year the $100 you earn in interest is added to your capital and during the second year you earn interest, not on your original $1,000, but on $1,100: your original capital plus the interest earned during the first year. The amount you have accumulated at the end of any given year is the amount you had invested at the beginning of that year multiplied by 1.1. Using annual compound interest your original $1,000 becomes $45,259 at the end of 40 years.

If you compound semiannually, you will have at the end of any six-month period what you had at the beginning of that period multiplied by 1.05 (if you get 10 percent interest per year you only get 5 percent interest per half-year). By compounding semiannually you will accumulate $49,561 in 40 years.

If the period of compounding is made smaller and smaller (so that you are compounding more and more frequently) the amount accumulated gets larger and larger but never exceeds an upper limit; the most you can ever accumulate in forty years, no matter how often you compound, is $54,598. To understand why this is so we need to examine the notion of *Continuous Compounding*.

Continuous Compounding

If you invest $1,000 for two years and use 10 percent simple interest, you will accumulate $1,200. If you compound annually, you will have $1,210 in two years, an *under*whelming difference of less than 1 percent. The fact is, however, that what appears to be a minor difference becomes a major one to the long-term investor.

Let's reconsider these computations more carefully. Starting with principal of P dollars and compounding annually at i percent you have $P(1 + i/100)$ or $P + P(i/100)$ at the end of a year. P is your original capital and $P(i/100)$ is the interest you accrue. To find your capital at the end of the second year, multiply the capital you had at the end of the first year, $P(1 + i/100)$, by $(1 + i/100)$ once again to obtain $P(1 + i/100)^2$. To calculate your new principal at the end of the Nth year, multiply the capital you had at the end of the $(N-1)$st year by $(1 + i/100)$ to obtain $P(1 + i/100)^N$.

So, for example, at 6 percent annual interest and with an initial $100 you would have $100(1 + 0.06)^{12} = 201.22 at the end of 12 years. If you compound more often than once a year the interest that you earn increases.

Suppose you start with the same $100 and interest at 6 percent and you lend your money for the same 12 years but the interest is compounded semi-annually. Since you get 6 percent interest for a year you only get 3 percent interest for six months (the compounding period) and there are 24 compounding periods. In one year you would have $100(1 + 0.03)^2 = 106.09$. This is $0.09 more than you would receive if you compounded annually. The $0.09 is 3 percent of $3, which is six months of interest on the $3 interest earned during the first six months. At the end of 12 years (24 six-month compounding periods) you would have $100(1 + 0.03)^{24} = 203.28—a modest improvement over the $201.22 from annual compounding.

To better understand the compounding process observe in Table 2.2 how $1,000 grows, quarter to quarter, if compounding occurs at 8 percent compounded quarterly (that is, 2 percent every three months).

While the effect of compound interest is impressive over longer periods, at a certain point reducing the length of the compounding period

| | Capital at the | Interest | Capital at |
| Period | Beginning | Earned During | the End of |
in Months	of the Period	the Period	the Period
0 to 3	$1,000.00	$20.00	$1,020.00
3 to 6	$1,020.00	$20.40	$1,040.40
6 to 9	$1,040.40	$20.81	$1,061.21
9 to 12	$1,061.21	$21.22	$1,082.43
12 to 15	$1,082.43	$21.65	$1,104.08
15 to 18	$1,104.08	$22.08	$1,126.16
18 to 21	$1,126.16	$22.53	$1,148.69
21 to 24	$1,148.69	$22.97	$1,171.66

TABLE 2.2

yields diminishing returns. Ads in the newspapers may proclaim that interest is "compounded daily" but the difference between interest earned with compounding every three months and compounding every day is really very small.

If you start with $1,000, a nominal 8 percent annual rate, and compound annually, your final capital will be $1,080 after one year. If you compound four times a year, you will have $(1.02)^4 \times \$1,000$ or $1,082.43 after one year—an increase of $2.43 or about 3 percent. If you compound 365 times a year you will have $(1 + 0.08/365)^{365} \times \$1,000$ or $1,083.28—an increase in interest earned of only $0.85 or about 1 percent over quarterly compounding.

If you fix both the duration of the loan and the nominal interest rate but compound more frequently, you will earn more and more interest, which suggests the following question: can you get rich by simply shortening the period you use to compound interest?

Although you earn more interest when you compound more often the interest earned rapidly levels off. As mentioned earlier, if you receive j percent interest in a certain period of time, $1 will become $\$(1 + j)^M$ in M periods of time. If you compound N times a year for M years at a nominal i percent interest rate the interest rate for each period of compounding will be i/N and there will be MN periods of compounding so that after M years

	Simple Interest	Semiannual	Quarterly	Monthly	Daily
TABLE 2.3					
Starting Amount	$100.00	$100.00	$100.00	$100.00	$100.00
Jan (End of Month)				$100.83	$100.84
Feb				$101.67	$101.69
Mar			$102.50	$102.52	$102.54
Apr				$103.38	$103.40
May				$104.24	$104.27
Jun		$105.00	$105.06	$105.11	$105.14
Jul				$105.98	$106.02
Aug				$106.86	$106.91
Sep			$107.69	$107.75	$107.81
Oct				$108.65	$108.71
Nov				$109.56	$109.63
Dec	$110.00	$110.25	$110.38	$110.47	$110.55

Growth Due to Compound Interest at 10 Percent During One Year

your \$1 will become $\$(1 + i/100N)^{MN}$. (The *nominal* rate of interest is lower than the *actual* rate of interest. If you invested \$100 at a nominal interest rate of 12 percent compounded quarterly, you would receive 3 percent interest in each quarter and your \$100 would become \$112.55, that is, your actual interest rate would be 12.55 percent.)

As N gets larger, $\$(1 + i/100N)^{MN}$ gets larger as well and gets arbitrarily close to, but never exceeds $\$e^{iM/100}$, where e is a well-known mathematical constant equal to 2.718…, the base of natural logarithms.

The fact that interest earnings keep increasing as the compounding period gets smaller may seem to contradict the fact that the interest earned is bounded. But the contradiction is an illusion.

If you imagine someone earning a dollar one day, half a dollar the next, and each day earning half the amount earned the day before, it is clear that the longer he works, the more money he earns. But it is also clear that he will never earn more than a total of \$2 no matter how long he works.

Returning to an earlier example where the nominal interest rate was 8 percent, the amount that $1,000 will become in one year will never be more than $1,000 e $^{0.08}$ = $1,083.29 no matter how often interest is compounded.

Although frequency of compounding is not important for short investments at low interest rates, for longer investments at high interest rates it is critical. We just observed that at a nominal 8 percent annual interest you get $82.43 rather than $80 when you change from annual to quarterly compounding—a 3-percent increase on the interest earned. But suppose your $1,000 was invested for forty years rather than one year and the nominal interest rate was 12 percent rather than 8 percent. Compounded annually, you would have $93,051 after forty years, but you would have $113,229 if you compounded quarterly—an increase of about 22 percent in the interest earned. If you increase the frequency of compounding sufficiently, you can raise the amount accumulated to as much as $121,510.

Calculating Interest—Another Common Error

Very often simple interest and compound interest are confused and this leads to errors that are a little subtler than the one we made in our opening dishwasher example. Take the following extreme example: Your neighborhood bookie offers to lend you $1,000 but charges you 1 percent daily, compounded daily. You keep the money for a year and then repay it with a lump sum. How much do you think you would have to repay?

Making an "approximate," calculation you might be tempted to think "1 percent a day equals 365 percent a year," so you would repay $3,650 in interest plus the $1,000 you borrowed: a total of $4,650. In fact, you would repay more than eight times that: $37,783. In an article in the *New York Times* (June 18, 1999, p. 28) about usurious loans, Peter T. Kilborn speaks of lenders who charge $33 for a two-week loan of $100 and equates this to an annual rate of (33 percent) (26) = 858 percent (since there are 26 two-week periods in a year). In fact the actual annual interest rate charged is $[(1.33)^{26} - 1] \times (100$ percent$)$, or 165,913 percent—a somewhat higher rate, to put it

	TABLE 2.4	
Interest Rate	N (the Predicted Doubling Time)	Actual amount $1 will become in N years
1%	70	$2.0068
2%	35	$1.9999
5%	14	$1.9799
7%	10	$1.9672
10%	7	$1.9487

mildly. So when your credit card company charges you 2 percent a month your real annual interest rate is 26.8 percent (not 24 percent); and if the inflation rate is 1 percent per month the true annual inflation rate is 12.68 percent, not 12 percent.

Doubling Your Money—How Long Does It Take?

There is a simple formula that is useful for certain interest calculations. If you invest your money at an annual i percent interest rate it will take approximately $70/i$ years for your money to double. If, for example, you deposit money at 8 percent interest, it will double in approximately $70/8 = 8.75$ years. (It will double actually in about 9.006 years.) The formula is only an approximation but is accurate for interest rates that do not exceed 10 percent. The accuracy of the formula is reflected in Table 2.4 (it is assumed that you start with a principal of $1).

Let's see how the formula is applied in practice. Suppose a house is purchased for $200,000 and sold twenty years later for $400,000. How could you approximate the annual rate of growth? Since the doubling time was twenty years the approximate rate of growth is 70/20 percent or $3^{1}/_{2}$ percent per year. (The actual rate of growth is about 3.526 percent.)

If you are unfamiliar with interest rates, you may be startled by the rate at which capital grows over long periods of time—even at modest annual growth rates. Suppose a house was bought in 1900 for $5,000 and sold in 1984 for $320,000—64 times the original purchase price. It would require a 75 percent simple interest rate for $5,000 to grow to $320,000 in 84 years.

But with compounding we have a very different story. Since the value of the house increased by a factor of 64 (= 2^6) it doubled six times in 84 years, which means the value of the house doubled every 84/6 = 14 years. (So at a constant growth rate it would have been worth $10,000 in 1914, $20,000 in 1928, and finally $320,000 in 1984). Doubling every 14 years means that the annual interest rate is only about 5 percent.

As a final illustration assume that Manhattan Island, bought in 1626 for $24, was sold in 1986 for $25,000,000,000,000—about 1,040,000,000,000 times its purchase price. How can we estimate the annual rate at which the price increased?

By solving 2^N = 1,040,000,000,000 we find that N is about 40 so that in 360 years your capital doubled 40 times. The number of years it takes for your capital to double once is about 360/40 = 9 years. 70/9 = 7.8 so the approximate annual growth rate is 7.8 percent. (The actual growth rate, given our hypothetical numbers, is about 8 percent.)

The justification for the formula is simple enough—given an interest rate i we seek an N such that $(1 + i)^N$ = 2. This can be restated as N $\text{Ln}(1 + i)$ = $\text{Ln}\,2$ where Ln is the natural logarithm. If i is small then $\text{Ln}(1 + i)$ is approximately i and $\text{Ln}\,2$ is approximately 0.70, so Ni = 0.70 (approximately).

We will complete this chapter by considering the role that interest plays in a variety of different applications—Social Security, lotteries, perpetuities, and deferred taxes.

Social Security—Should You Take the Money and Run?

Until recently, a person who retired before age 70 was faced with a dilemma—whether to start collecting Social Security benefits at once and receive a fixed annual payment for the rest of his life or wait until a later date and receive a higher annual benefit. Now the same choice is available for those over 65 whether they are employed or not. (For the sake of simplicity

	Payments Start at 62 Option (a)	Payments Start at 65 Option (b)
Age 62	$ 0	$ 0
Age 65	$ 24,000	$ 0
Age 68	$ 48,000	$ 30,000
Age 71	$ 72,000	$ 60,000
Age 74	$ 96,000	$ 90,000
Age 77	$120,000	$120,000

we will ignore cost of living increases and deal with the real interest rate—
the amount that the interest rate exceeds the inflation rate). Does a retiree
do better to wait or collect immediately? To fix our ideas we will make certain assumptions about the rules governing social security payments (which,
in any case, are constantly changing).

Suppose that at age 62 you have two choices:

(a) You can receive $8,000 for the current year and $8,000 each year
    thereafter for the rest of your life.

(b) You can receive $10,000 each year for life with the first payment
    at age 65.

Clearly the longer you live, the more attractive the delayed payments.
We will try to determine how long you must live so that the value of both
streams of payments are the same.

If you are paid from the beginning of the first year until the beginning
of the nth year ($n > 2$) and evaluate your payments at the end of the nth year
you would obtain a total of $8,000n$ using (a) and you would get $10,000
($n - 3$) using (b). The amounts are equal when $n = 15$, that is, when you are
77 years old. This is illustrated in Table 2.5.

The entries in the table indicate the total amount received by recipients in each plan at three-year intervals assuming they are still alive.

If you live to age 77 you will have obtained the same amount of money with options (a) and (b) but the two streams of payment are not equivalent. Since a dollar received today is worth more than one received tomorrow the value of payments under (a) is worth more than that under (b) because payments were received earlier. If the annual rate of interest is i and interest is taken into account, the value of the payments under (a) is $\$8,000[(1 + i)^{n+1} - (1 + i)]/i$ at the end of the nth year and under (b) the value of the payments is $\$10,000[(1 + i)^{n-2} - (1 + i)]/i$ (assuming you live at least three years). If $i = 2$ percent, you must wait until age 80 for plan (b) to be better than (a), if $i = 4$ percent you must wait until age 83 for plan (b) to be better than (a), if $i = 6$ percent, the critical age is 90 and if $i = 8$ percent the $\$8,000$ payment stream will be superior to the delayed $\$10,000$ payment stream no matter how long you live!

Consider what happens when $i = 8$ percent. Suppose you take the $\$8,000$ payments and save them, assuming they earn 8 percent interest. At the beginning of the third year your first three payments will be worth $\$8,000[(1.08)^2 + (1.08) + 1] = \$25,971$. If you leave this amount in the bank and simply collect the interest, you will receive more than $\$2,077$ a year for the rest of your life which more than compensates for the $\$2,000$ difference in payments between plans (a) and (b) and also leaves a principal of more than $\$25,000$ at your death.

When the interest is exactly 7.7 percent you would have to live to age 143 for the $\$10,000$ payment option to be superior.

The point is this: when you try to evaluate a stream of payments over an extended period of time you grossly distort your conclusions if you ignore interest. If you overlook the effect of interest on your Social Security calculations, you need only live to age 77 for delayed payments to be worthwhile. But even at moderate rates of interest you are unlikely to live long enough for delayed payments to be worthwhile and when interest rates are *high* it is better to take the immediate smaller payments no matter how long you live.

TABLE 2.6						
Annual Interest Rate	2%	4%	6%	8%	10%	12%
Discounted Value of Million Dollar Prize	$834,00	$707,000	$608,000	$530,000	$468,000	$418,000

When Is a Million Dollar Lottery Worth Half That?

Various aspects of everyday life seem unrelated to interest but when you look closer, you find interest playing an important role. Lottery promoters might describe their first prize as a "million-dollar payoff" but on closer inspection the million dollars turns out to be $50,000 a year paid at the beginning of each year of the next twenty years. While you do eventually get your million dollars its value is reduced considerably when you take into account the delay in payment. If the interest rate is i, the discounted, or present, value of your payments when you first win the lottery is the sum of the geometric series $50,000 $[1 + 1/(1 + i) + 1/(1 + i)^2 + \ldots + 1/(1 + i)^{19}]$ = $50,000 $\{[1 - 1/(1 + i)^{20}] / 1 - 1/(1 + i)]\}$; this discounted value may be considerably less than a million dollars depending upon the prevailing interest rate. (A geometric series is a sum in the form $S = a + ar^2 + \ldots + ar^{n-1}$ in which the ratio of any term to its predecessor is constant: r.) In this case $S = (a - ar^n)/(1 - r)$. If r is less than one in magnitude and n is large, S is approximately equal to $a/(1 - r)$.

The actual value of the "million dollar" prize is shown in Table 2.6 for various assumed interest rates. When interest is high (9 percent or more) the actual value of the payments you receive is less than half the amount that was advertised.

A Perpetuity—An Infinite Number of Payments With Only Finite Value

To see how critical interest rates can be in financial calculations imagine that the lottery winner receives, not just twenty $50,000 payments, but payments that continue forever. (After your death payments continue to be paid to your estate—bonds like this really do exist and are called perpetuities or consols.) Since the payments are constant and the number of payments is clearly infinite the amount that is paid out is clearly infinite as well. You might guess that even after you take the erosion due to interest discounting into account the actual value of the payments would be infinite as well, but you would be mistaken; by discounting with interest you add an infinite number of equal payments and get a finite sum. (Imagine you put $10,000 in the bank at 10 percent annual interest and withdraw $1,000 at the end of the year. Since you are only withdrawing earned interest and none of the principal you can do this perpetually. In effect, your present $10,000 is worth an eternal flow of $1,000 annual payments starting a year from now.)

The value of the sum of the payments with interest rate i is $\$50,000[1 + 1/(1 + i) + 1/(1 + i)^2 + 1/(1 + i)^3 + \ldots] = \$50,000[1 + 1/i]$.

At 6 percent interest the infinite number of payments are only worth $883,333 and at 10 percent interest the payments are only worth $550,000. Once again, this is not too surprising. If you have $550,000 and a bank is willing to pay you 10 percent interest, you could withdraw $50,000 at the start of the year and let the remaining $500,000 earn interest for the rest of the year. At year's end you have earned the $50,000 in interest that you withdrew earlier and you can start all over again the following year with the same capital—that is, you can pay out $50,000 per year and never run out of money.

Death and Taxes Are Inevitable—But Defer the Taxes

If you take advantage of current tax laws, you may have the opportunity of postponing the taxes on some of your interest earnings by setting them aside in special investment accounts. The pros and cons of doing so are intimately involved with interest considerations. Eventually, when these earnings are actually withdrawn from the special accounts, the taxes must be paid. The alternative to such an investment is to pay the taxes each year on the interest as it is earned.

For the sake of simplicity, assume you always pay taxes on interest earned at one fixed rate. It might seem at first that the two investment procedures are equivalent since tax is paid on interest earnings exactly once in both cases, but actually there is a substantial advantage to postponing your tax payments as long as you can. This is because the money that you would have paid in taxes (if you paid annually) remains in your account and earns additional interest, and over long periods of time this can be very significant.

A simple illustration: Suppose you invest $100, the interest rate is 10 percent and you pay 20 percent in taxes on the interest you earn. If you defer taxes for two years, you will have $110 after one year and $121 after two years. Since you earned a total $21 in interest you must pay 20 percent of that or $4.20 in taxes and are left with $116.80.

If you pay taxes each year, you will earn $10 the first year and will be taxed $2, ending the year with $108. In the second year your $108 will earn $10.80 and you will be taxed $2.16, leaving you with a net amount of $116.64, which is somewhat less than the $116.80 you earned with the tax delay. The $.16 difference between the earnings is one year's interest on $2 in taxes (paid in one case but not in the other) less 20 percent in taxes. The 16-cent difference may not seem like much but over longer periods and with larger sums the difference matters.

Suppose you invest $1,000 and pay taxes on the interest each year for a period of forty years. If the marginal tax rate is 25 percent and the interest rate is 8 percent your money grows at an effective rate of (0.75) (8 percent) = 6 percent after taxes. The $1,000 payment will be worth $1,000 $(1.06)^{40}$ = $10,285.72.

If no taxes are deducted for forty years, you will have $1,000 $(1.08)^{40}$ = $21,724.52 before taxes are deducted. The interest you earned, $20,724.52, is taxable at 25 percent so you have after taxes $21,724.52 − (0.25) ($20,724.52) = $16,543.39 which is 60 percent more than you had when you paid all you taxes annually. (For the math majors, suppose T is the single tax rate and i is the interest rate. After n years, if you pay your taxes each year, $1 becomes $[1 + i(1-T)]^n$; if taxes are deferred n years, $1 becomes $[(1 + i)^n(1 - T) + T]$.

Averaging Percentage Returns on Investments

A final example: Suppose you have the opportunity to make a sequence of speculative investments—in the theater, for example—in which you quadruple your money 10 percent of the time and break even the rest of the time. You always invest all the capital you have on whatever investment you are making and withdraw nothing. When you are successful you make a 300-percent profit and when you are unsuccessful you are at breakeven with a zero percent profit. What is your average profit?

Although you might guess that your average profit is (0.1) 300 percent + (0.9) 0 percent = 30 percent, it is in fact less than half that. To see why, let's first decide what we mean by "average." Suppose your daily earnings are $200, $250, $300, $350, and $400 during a particular week. Your average earnings would then be $300, which means that if you earned exactly $300 every day, you would get the same weekly earnings as the weekly earnings that you actually received: $1,500.

Using this same approach you might ask, "If I made the same percentage profit each year what percentage would that have to be so that my long term profits would equal what I would actually earn from nine failures and one success?"

When you succeed you multiply your capital by four and when you fail you multiply your capital by one so the effect of nine failures and one success is to multiply your capital by $(1)^9 (4)^1 = 4$. If you had the same return, i, every time and you made ten investments you would have after ten investments $(1 + i)^{10}$. Setting $(1 + i)^{10} = 4$, we find $i = 4^{0.1} - 1 = 14.9$ percent. If you actually received a 30 percent return on each investment (as you might have thought initially) you would have $[(1.3)^{10} - 1] = 12.8$ times your original capital, on average, after ten investments (rather than the four times your original capital that you actually received).

If, as we said earlier, interest rates significantly effect our standard of living, it is important that the choices that we make be informed. Unfortunately, both the numbers and the concepts can be deceptive. We should know that the "trivial" 2 percent interest per month we pay on our credit card reduces the buying power of $100 to less than $74 in one year. We should know that the $100 that we borrow from a loan shark at an apparently modest 5 percent per week would require us to repay $1,264 if we waited until the end of a year. In this chapter we have shown that you can be paying twice the interest you think you are paying and introduced you to the magic of compound interest growth. In Chapter 4 you will see how prudent choices can substantially lessen the amount of interest you pay on a mortgage.

Your insight into the workings of interest should enable you to calculate rapidly the time it will take to double your capital, decide whether to delay your Social Security payments, determine whether a "monster" lottery prize is as large as it purports to be, and see the advantages of delaying taxes on your earnings.

Chapter 3

Bonds

Test Your Intuition 3

When you buy a bond you pay for it at once; in return you receive a dividend of $D at the end of each of the next N consecutive years and at the end of the Nth year you receive an additional $1,000 payment as well.

{1} In the real bond market the interest you get on a riskless bond will vary with the maturity date at any one particular time. Usually, the longer the time to maturity, the higher the interest rate. But is it possible, in principle, to have a bond market in which all bonds have the same interest rate at any given time, whatever their maturity date (but an interest rate that might change over time)?

{2} You plan to buy a house T years from now so you buy a bond with a duration of T years (the duration of a bond is the average time that elapses between the time you buy the bond and the time you receive your repayments) and with an interest rate of i percent. You plan to reinvest all your dividends at the prevailing rate as you receive them and then to withdraw your dividends and sell your bond in T years to buy the house. Right after you buy your house interest goes to j percent and never changes. If $j > i$, your bond is worth less when you sell it but your dividends earn more when you reinvest them; if $j < i$, your bond will be worth more when you sell it but your dividends will earn less. Should you be hoping for an increase, or a decrease, in the interest rate?

{3} You own a 30-year bond that pays a $50 dividend. If interest rates increase from 7 to 8 percent the value of the bond will decrease by 11.9 percent. Guess how much your bond will drop if it were a zero-coupon bond (a bond paying no dividends) rather than one paying an annual $50 dividend.

Answers on page 188.

In case you are someone who likes to skip around, before you read this chapter you should be familiar with interest rates, the subject of the preceding chapter.

What Is a Bond and What Should
You Know Before You Buy One?

When you buy a bond you are really making a loan. You—the bondholder—first pay for the bond (make the loan) and then you receive a stream of repayments from the issuer (that pays you interest on the loan and the loan itself). An issuer might be anyone needing money: the U. S. government or one of its agencies, a state authority building a new freeway, or a business in need of cash. A bond buyer might be a person or corporation interested in earning money on capital. You can think of the cost of a bond as the loan principal but the rate of interest (called the yield to maturity for bonds) is harder to calculate because there are many, not just one, repayments.

Although bonds come in many shapes and sizes we will focus on just a few typical examples. We will generally consider $1,000 bonds (many, but not all, bonds are $1,000 bonds), that pay the same dividend (or coupon) at the end of every year until the bond matures. We will also generally assume, unless we state otherwise, that the repayments will be made as scheduled (a real bond, of course, may not be repaid on time or repaid at all). For the sake of simplicity we will ignore such complications as early call provisions and tax considerations.

What should you know before you buy a bond? You certainly should know the cost of the bond and the schedule of repayments—from this information you can derive the effective interest rate. You should also be aware of the prevailing interest rate and compare it to the interest rate that the bond actually pays. (And for a real bond you should have some idea of the risk of default, as well. Bond ratings are in the public record and may be easily obtained). You should also know how future changes in the prevailing interest rate will affect the value of your bond if you sell it before it matures.

From the chapter on interest you know that the worth of future repayments can not be taken at their face value but must be discounted with interest. This is a consequence of the fact that a dollar received today is worth more than the dollar received tomorrow. So a bond that pays you $100 at the end of every year perpetually when interest is 10 percent does not have an infinite value (the actual sum of all future payments); such a bond is only worth $1,000 after the payments are discounted with interest.

A Typical Bond

The bonds we will discuss will generally have the same basic structure. Typically, you might pay $862.35 today for a $1,000 face-value bond in a 6 percent world. In return for this payment you would receive thirty $50 dividends at the end of each of the next thirty consecutive years with the first dividend coming one year after the purchase date. In addition, you would receive a $1,000 payment at the end of the thirtieth year. If you calculated all your repayments (including the $1,000), and then discounted them at 6 percent to the purchase date and summed them, you would have your purchase price.

In general, one year after you buy a bond you get your first dividend of $D and you continue to get the same annual dividend of $D for N consecutive years. At the end of the Nth year you get a lump sum of $1,000 in addition to your last dividend of $D (see Figure 3.1). The cost of the bond is equivalent to all the repayments after they have been discounted with interest; the rate of interest earned is also called the yield to maturity.

Year received	0	1	2	3...	...N
Amount received		$D	$D	$D...	...($1,000 + D)

FIGURE 3.1

Calculating the Value of a Bond

The value of a bond is determined by three variables: N (the number of years to maturity), i (the prevailing interest rate) and D (the size of the dividend or coupon). It is convenient to separate the revenue from a bond into two parts: the dividends that you receive periodically and the final $1,000 payment.

If the interest rate is i, a payment of D, k years from now, is worth $D(1+i)^{-k}$ now. The present value (at the time of purchase) of the N dividend payments is therefore the sum of the following geometric progression: $D\{[1/(1 + i)] + [1/(1 + i)^2] + \ldots + [1/(1 + i)^N]\} = [D/i][1 - (1 + i)^{-N}]$. (See Chapter 2 for the formula for the sum of a geometric progression.) The last $1,000 payment is worth (when you buy the bond) $1,000(1 + i)^{-N}$.

The value of your bond—the sum of the values of the dividends and final payment—is given by the formula: $V = (D/i) + [1,000 - (D/i)](1 + i)^{-N}$, where D is your annual dividend, i is the yield to maturity (or interest rate), N is the number of years to maturity and V is the bond's value on the date it is purchased.

A few special cases may make the formula more meaningful. Suppose that N is very large; in that case $(1 + i)^{-N}$ would be very small and V would be approximately equal to (D/i). If the bond we described earlier paid $50 dividends forever (so there would be no final payment), its value would be $50/0.06 = $833.33 when $i = 0.06$. If the bond we described earlier paid no dividends (so $D = 0$) but had a final $1,000 payment in 30 years, its value now would be $1,000(1.06)^{-30} = $174.11.

How Bond Prices Change Over Time

Bonds often have to be sold unexpectedly so bondholders should be concerned about what their bond is worth even before they mature. When the prevailing interest rate changes, the bond's value will change as well. But, even if interest rates remain constant, there is a variation in bond prices throughout their life and the variation is predictable. In fact if interest rates do not change, a bond's value will have one of three life histories. Assume, for simplicity, that we have a 30-year bond with interest at 5 percent.

(a) D = $1,000 i. In this case the dividend is exactly the same as the interest you would earn on $1,000. If you put your $1,000 in a bank at 5 percent you would generate $50 a year in interest at the end of every year and you would have your $1,000 at the end of the year, as well. As long as interest remains the same your capital would never change. In effect, your bond is generating $50 a year in exactly the same way the bank did and the value of your bond remains constant—$1,000.

If you formally substitute D = 50 and i = 5 percent into the formula, $[1,000 - D/i] = 0$ and therefore $V = D/i = 50/0.05 = 1,000$; the bond's value is initially $1,000 and remains $1,000.

(b) D < $1,000 i = 50. In this case your dividends are less than you could earn from a bank earning i percent interest although you still get the $1,000 at maturity. Because of the lower dividends this bond is worth less than $1,000—the value of the bond described in (a). As time passes, the years to maturity, N, decreases and the second term gets larger (the first term is constant). At maturity N = 0 and the value of the bond is 1,000. In short, a bond that pays dividends lower than the prevailing rate of interest has an initial value less than $1,000 but keeps increasing in value until it reaches $1,000 at maturity.

(c) When the dividend is larger than the interest paid on $1,000, that is, D > $1,000 i = 50, the initial value of the bond is more than $1,000 but as you approach maturity the bond's value approaches $1,000 from above.

The Effect of Duration, Years to Maturity, and Interest on Bond Prices

The value of a bond is directly related to the prevailing interest rate. When the interest rate falls, bond values rise; when the interest rate rises, bond values fall. But the change in value of a bond that results from a change in the interest rate can vary considerably from bond to bond.

Consider a variety of bonds which mature anywhere from 5 to 800 years from now, which pay dividends anywhere from $0 to $120 per year. Also suppose that the prevailing interest rate can vary from 1 percent to 9 percent. If the prevailing interest rate increases by 1 percent the bond will inevitably decrease in value but it may do so by as little as 4.1 percent or by as much as 50 percent. How is this possible? To determine the sensitivity of a bond to a 1 percent change in the interest rate you must consider the years to maturity, the size of the dividend and the prevailing interest rate at the time the bond was bought.

Years to Maturity

Other things being equal, the value of long-term bonds will be more sensitive to a change of 1 percent in the interest rate than short-term bonds. If you buy a five-year bond when interest is at 5 percent and the interest rate rises to 10 percent, your bond decreases in value; the money you spend for a bond in a 5 percent world entitles you to higher dividends when you enter the 10 percent world. When you buy your five-year bond in the 5 percent world you commit yourself to lower dividends (lower than those offered in a 10 percent world) for five years, so the bond's value naturally drops; it would drop even more if you bought a ten-year bond and your commitment to lower rates was even longer. (Your gains would be greater as well if the interest rate went in the other direction.) So the longer the term of a bond the greater your commitment and the more sensitive the bond to interest rate changes. If the prevailing interest rate goes from 5 to 6 percent the value of a five-year bond paying $50/year in dividends would go from $1,000 to $958—a drop of 4.2 percent. The price of an 800-year bond paying the same dividend would go from $1,000 to $833—a drop of 16.7 percent.

Dividend Size

Other things being equal, the smaller the dividend, the greater a bond's sensitivity to interest rate changes. If the prevailing interest rate goes from 5 percent to 6 percent, the price of a 30-year bond with no dividends would go from $231 to $174—a drop of 24.8 percent. The price of a similar bond paying $120/year dividends under the same conditions would go from $2,076 to $1,826—a drop of 12.1 percent.

The Prevailing Interest Rate

Other things being equal, the lower the prevailing interest rate, the greater the change in a bond's value when there is a change of 1 percent in the prevailing interest rate. Suppose a 30-year bond pays a $50 dividend. If interest rates rose from 1 percent to 2 percent, the value of the bond would go from $2,032 to $1,672—a drop of 17.7 percent. If interest rates went up from 8 to 9 percent, the bond's value would go from $662 to $589—a drop of 11.1 percent.

A Precise Definition of Duration

There is a particular characteristic of a bond that indicates how sensitive it is to changes in the interest rate: the longer the repayment period, the greater the bond's sensitivity to changes in the interest rate. But exactly how long is the repayment period when there are multiple repayments? Some dividend payments are made as early as the first year, with the last dividend payment and the final $1,000 payment coming at maturity. What we need is a single number that reflects the average repayment time. This average would take into account the size of each payment, its discounted value, and the time that the payment was made.

If you borrow $100 today and repay $120 in exactly one year, the repayment period is clearly one year and the annual interest is 20 percent. But suppose you repay $60 after six months and another $60 at the end of a year. How long is the repayment period in that case? You might be tempted to average one year and six months and say nine months but this would be incorrect. The value of the discounted later payment is less than the value of the discounted earlier payment, and so the earlier payment should be weighted more heavily. Think of it this way: the annual interest rate is 27.839 percent (so that the interest rate for six months is 13.066 percent) and you borrow $53.066 for six months and $46.934 for a year (a total of $100). At the end of six months you repay $53.066 × (1.13066) = $60 and at the end of a year you repay $46.934 × (1.27839) = $60. The duration, the average repayment time, would therefore be closer to six months than one year since you actually borrowed more money for the six-month loan ($53,066) than you did for the one-year loan ($46,934).

Associated with each bond is a schedule of payments that occur between the time of purchase and maturity. The average time that passes between the time you buy the bond and the time you receive your repayment (weighted by the size of the payment and discounted to the purchase date) is called the duration of the bond. The duration is given by the formula

$$\text{Duration} = \frac{P_1 \, T_1 \, (1+i)^{-T_1} + P_2 \, T_2 \, (1+i)^{-T_2} + \ldots\ldots + P_N \, T_N \, (1+i)^{-T_N}}{P_1 \, (1+i)^{-T_1} + P_2 \, (1+i)^{-T_2} + \ldots\ldots + P_N \, (1+i)^{-T_N}}$$

where P_j is the jth payment made after T_j years and i is the interest rate.

If, for example, you have a two-year bond that pays $50 at the end of the first year and $1,050 at the end of the second year when interest is 10 percent the duration would be

$$\text{Duration} = \frac{(50)(1)(1.1)^{-1} + (1,050)(2)(1.1)^{-2}}{(50)(1.1)^{-1} + (1,050)(1.1)^{-2}} = 1.9502$$

This reflects the fact that most of the repayment is made late in the second year.

Returning to our earlier example where you borrowed $100 and made a $60 payment after six months and another $60 payment after a year, you would have $P_1 = P_2 = \$60$, $T_1 = 6$ months, $T_2 = 12$ months and $i = 0.13066$ it turns out duration = 8.816 months: just under nine months.

This is the weighted average of the durations of the two hypothetical loans: [(53.066)(6 months) + (46.934)(12 months)] / 100 = 8.815 months.

If you want to know how a characteristic of a bond affects its sensitivity to interest rate changes, it isn't really necessary to calculate the duration exactly. It is enough to know that the longer the bond's duration, the greater its sensitivity. By observing how this characteristic effects the duration you can deduce the sensitivity fairly accurately. When dividends are small, the last $1,000 payment is relatively large, so the duration, the average repayment period, is large and the sensitivity to interest rate changes is high. Similarly, if two similar bonds have different maturities, the bond maturing later will have a greater duration and be more sensitive to interest rate changes. And finally, if two bonds are identical but one is bought when the prevailing interest rate is higher, the early payments on the bond with the higher interest rate will be more significant since the late payments will be discounted more. Thus the duration will be less, and there will be less sensitivity to interest rate changes.

Bond Price Sensitivity—A Formula

If you know the duration of a bond, R, the value of a bond, V, and the prevailing interest rate, i, there is an approximate formula which reflects a bond's sensitivity to the interest rate: $(\Delta V/V) = -[R/(1 + i)] (\Delta i)$, where ΔV is the change induced in the bond price by a small change in the interest rate, Δi.

It follows immediately from this formula that if the interest rate changes, bond values change in the opposite direction; the greater the duration, the greater the proportionate change in the bond's value.

In Table 3.1, five different bonds are listed with various maturities, interest rates, and dividends. For each of them we show by what percentage the value of the bond would change if interest increased by 1 percent.

If you bought the 30-year bond in the first column (annual dividend $50) when interest was at 1 percent and sold it immediately when interest rose to 2 percent, the value of the bond would decrease by 17.7 percent. If you bought the same bond when interest was at 9 percent and then increased by 1 percent to 10 percent, the bond's value would decrease by 10.3 percent. At higher interest rates the duration is shorter and the value of

TABLE 3.1

Yield to Maturity	N = 30, D = 50		N = 5, D = 50		N = 800, D = 50		N = 30, D = 0		N = 30, D = 120	
	Bond Value	% of Value Drop	Bond Value	% of Value Drop	Bond Value	% of Value Drop	Bond Value	% of Value Drop	Bond Value	% of Value Drop
1%	2,032	17.7	1,194	4.4	4,999	50.0	742	25.6	3,839	15.6
2%	1,672	16.7	1,141	4.4	2,500	33.3	552	25.4	3,240	14.7
3%	1,392	15.7	1,092	4.3	1,667	25.0	412	25.2	2,764	13.8
4%	1,173	14.7	1,045	4.3	1,250	20.0	308	25.0	2,383	12.9
5%	1,000	13.8	1,000	4.2	1,000	16.7	231	24.8	2,076	12.1
6%	862	12.8	958	4.2	833	14.3	174	24.5	1,826	11.1
7%	752	11.9	920	4.1	714	12.5	131	24.4	1,620	10.5
8%	662	11.1	880	4.1	625	11.1	99	24.2	1,450	9.8
9%	589	10.3	844	4.1	556	10.0	75	24.0	1,300	9.1

the bond is more stable. When the duration is much shorter (column 2) the value of the bond is more stable. When the duration of the bond is much longer (column 3), the value of the bond is less stable. Zero-coupon bonds (column 4) have a long duration and are therefore sensitive to interest changes and the large dividends in column 5 reduce the sensitivity of the bond to interest rate changes.

Junk Bonds

To this point we have assumed that bonds do not default. The fact is that they do. Bonds that are issued by shaky enterprises with a higher probability of defaulting are called junk bonds. For obvious reasons they offer a higher interest rates than other bonds.

Given the choice of a bond that has no risk of default and a higher yielding but riskier bond—what factors should affect your choice? The complete answer is very complex and involves your attitude toward risk, the probability of default, when you think the default is likely to occur, among other things. To simplify the problem we will make certain simplifying assumptions.

Suppose a stockholder wants to maximize the expected value of his profits. Suppose also that each junk bond has a certain probability of

TABLE 3.2

Type of Bond	Dividend	Prevailing Interest Rate	Term of the Bond	Value of the Bond	Probability of Default Making Bonds Equivalent
Riskless	$50	2%	30 years	$1,672	
Junk	$70	2%	30 years	$2,120 if no default	0.367
				$899 if default	
Riskless	$50	6%	30 years	$862	
Junk	$70	6%	30 years	$1,135 if no default	0.561
				$650 if default	
Riskless	$50	10%	30 years	$529	
Junk	$70	10%	30 years	$717 if no default	Junk bond better
				$532 if default	even if default certain.
Riskless	$50	2%	10 years	$1,269	
Junk	$70	2%	10 years	$1,449 if no default	0.161
				$330 if default	
Riskless	$50	6%	10 years	$926	
Junk	$70	6%	10 years	$1,074 if no default	0.190
				$295 if default	
Riskless	$50	10%	10 years	$693	
Junk	$70	10%	10 years	$816 if no default	0.223
				$265 if default	

defaulting; if it defaults, it will default precisely at midlife, paying dividends up to half the time to maturity but paying no dividends during the second half of the life of the bond and not paying the final $1,000 payment. If the junk bond does not default, it makes all obligatory payments. Under these assumptions what is the most opportune time to buy a junk bond? Do you do best when the prevailing interest rates are high or when they are low? Are short-term junk bonds more or less attractive than long-term junk bonds? Most important: what probability of default would make a riskless bond and a junk bond equally attractive?

Consider a simple example: assume that the prevailing interest is 6 percent and you have a choice of two 30-year bonds. The first, a riskless bond, pays $50/year in dividends; the second, a junk bond, pays $70 a year in dividends. The safe bond is worth $862; the junk bond is worth $1,135 if it does not default and $650 if it does. If you are not risk-averse but only concerned with your average return, you should prefer the junk bond if it's probability of defaulting is less than 0.562. To confirm this assume that the probability of defaulting is exactly 0.562, which means the probability of not defaulting is 0.438. On average, the junk bond would be worth $(0.562)(\$650) + (0.438)(\$1,135) = \$862$—the value of the riskless bond.

If you examine the data shown in Table 3.2 you can see how duration and the prevailing interest rate affect the attractiveness of junk bonds. All the junk bonds shown below pay dividends of $70 per year and the default-free riskless bonds pay $50 per year. By calculating the expected value of the riskless and junk bonds you will find that the probability of default of the junk bond will give both bonds the same expected value.

Examining the table, it is easy to see that high interest rates and long maturities make junk bonds more attractive. Notice that when interest is at 10 percent the junk bond is worth more than the riskless bond even if a default is certain. If you feel that high coupon bonds are reasonably secure in the short run (though somewhat doubtful in the long run) and if you are not averse to risk (that is, you measure the attraction of a bond by its average return), the existence of a larger dividend may adequately compensate you for the loss of half your dividend payments in the case of default; this is especially true when interest rates are high and the bonds have a long duration. Although the specific numbers about bond values and probabilities are based on a somewhat arbitrary assumption about when default will occur, the conclusions about the desirability of long-term, high interest junk bonds are generally true.

Can One Interest Rate Fit All?

Earlier we mentioned that the price of a bond is partially determined by the prevailing interest rate as though one interest rate fits all; but there are actually many different interest rates that are in effect at any one time, and these vary with a bond's duration. But could it be otherwise? Could there be a world in which, at any given time, there is one single interest rate that applies to all riskless bonds whatever their duration? (This rate would also apply to all bonds that are to be bought in the future as long as the commitment to buy is made now.) At other times there would also be a single interest rate but that rate might differ from the one in effect now. As it happens, with one trivial exception such a world could not exist. (The following example is perhaps for the more mathematically inclined; though all that is required to follow the argument is some algebraic agility and persistence.) If there were such a world, you could fashion an arbitrage scheme that would enable you to make a profit whenever interest rates changed, up or down, with no risk of ever taking a loss.

To construct such an arbitrage scheme we will make use of futures on zero-coupon bonds. A futures contract is an agreement by one party to buy, and another to sell, some commodity at some specific future time at a price fixed in advance. We will restrict ourselves to using only zero-coupon bonds. We assume you can buy and sell bonds at the same price in accordance with the prevailing interest rate. We also assume that it is possible to buy any fraction of a bond—in practice, if you are buying large quantities of bonds there is no difficulty closely approximating this arbitrage scheme.

Suppose that the interest rate is now at i percent and you commit yourself to buying a two-year bond one year from now. Since the commitment is made now the i percent interest rate governs and you will have to pay $[1000/(1 + i/100)^2]$. Suppose you also agree now to sell $[2/(1 + i/100)]$ one-year bonds a year from now for which you would receive $[2/(1 + i/100)]$ $[1000/(1 + i/100)]$ next year.

After one year suppose the prevailing interest rate becomes j percent and you undo both contracts; you sell the two-year bond and receive

[1000/(1 + j/100)^2] and buy [2 /(1 + i/100)] one-year bonds paying [2/(1 + i/100)] [1000/(1 + j/100)]. Everything you bought, you sold; everything you sold, you bought. If you subtract what you paid from what you received you will have

$$1000 \left(\frac{1}{(1+i/100)^2} - \frac{2}{(1 + i/100)(1+j/100)} + \frac{1}{(1+j/100)^2} \right) = 1000 \left(\frac{1}{(1+i/100)} - \frac{1}{(1+j/100)} \right)^2$$

and this is greater than zero if i is unequal to j. So if the interest changes, up or down, you make a profit and if interest remains unchanged you break even. We have constructed a riskless hedge.

One of the axioms of economics is that there cannot be an enduring, riskless hedge. Such a hedge, if adopted by everyone, would create a situation in which, in effect, people—everybody—would make a living by taking in each other's wash. The only world in which a duration-independent interest rate could exist would be one in which the interest rate did not change at all.

The Benevolence of (Some) Interest Rate Changes

The last topic in this chapter is of interest to someone who buys a bond and expects to hold it for some time. Changes in the interest rate are of concern to a long-term bondholder, but it is not clear whether you should be cheering for higher or lower interest rates. If after you buy a bond interest rates rise, the immediate effect is that the value of your bond will drop. On the other hand, the dividends you receive periodically can be reinvested at a higher rate of interest. If interest rates fall, long-term bondholders will still be ambivalent, but for the opposite reason: the value of the bond will increase but dividends will yield smaller returns in the future.

There is a theorem which partially resolves this ambivalence, at least under some special circumstances.

Suppose you plan to buy a house in T years. You buy a bond that has a duration of T years (the maturity date will therefore be more than T years from now) with an i percent yield to maturity. You intend to reinvest all your dividends as you receive them and T years after the bond purchase you sell the bond for whatever it is worth and cash in your invested dividends as well. Assuming that the interest rate never varies from i percent, it is a simple calculation to determine how much money you will accumulate to buy the house.

But suppose immediately after purchase the interest rate changes to j percent (and remains j percent thereafter). If $j > i$ your bond will be worth less (than it would if there were no change in the interest rate) when you sell it in T years, but the dividends will have been invested at a higher rate so it isn't immediately obvious whether you gain or lose. If $j < i$, the overall outcome will still be uncertain because the higher bond price T years from now will be balanced by the decrease in earnings from the dividend reinvestments.

The theorem states (for this specific behavior of the interest rate) that you will have at least as much money T years from now as you would have had if interest rates remained constant however the interest rate changes. If you sell when your bond reaches its duration the increased value of your bond will outweigh the lower dividend earnings when interest drops and the extra dividend earnings will outweigh the decreased value of your bond when interest rates rise. At worst, the interest rate will remain unchanged and you will have the amount originally anticipated.

Although our assumptions about the behavior of interest rates are artificial, the theorem does suggest that random changes in interest rates are likely to be benign but of course there is always a possibility you may be disappointed if interest rates are perverse. If, for example, interest rates drop below i percent immediately after the bond is bought and remain there for just under T years and then rise precipitately to above i percent the Tth year, you would have the worst of both worlds—lower reinvestment rates and a lower bond price.

Chapter 4

Mortgages

{1} Suppose you take out a 40-year mortgage for $100,000 at 16 percent interest and repay it in equal, monthly payments.

(a) Estimate the total amount of money that you will repay.
   i. $200,000   **ii.** $600,000   **iii.** $1,000,000

(b) Your monthly payments for the mortgage in (a) would be $1,248. If you change the 40-year mortgage to a 36-year mortgage, guess the amount that your monthly payments will be increased.
   i. $2   **ii.** $10   **iii.** $100

(c) Estimate how much money of the original loan you will owe after you have finished paying 36 of the 40 years of your mortgage.
   i. $ 2,000   **ii.** $4,000   **iii.** $45,000

(d) Estimate the fraction of your first monthly payment that is used to repay your mortgage (the remainder is paying interest on your loan).

{2} If you borrow $100,000 for 30 years at 8 percent interest you will pay $714 month. If you pay $X the first month and increase your monthly payments by 3 percent every year, your first year's monthly payment, $X, will be lower than the constant $714 payment by about
   i. 0.3 percent   **ii.** 10 percent   **iii.** 25 percent

Answers on page 188.

A Mortgage—The Mirror Image of a Bond

When you need money to buy a house it is customary to take out a mortgage. Taking out a mortgage is much the same as buying a bond, but in reverse. When you buy a bond you pay a large lump sum initially and then receive a stream of equal payments over time plus an additional lump sum when the bond matures. When you take out a mortgage you receive a large sum initially and in return you repay the loan with a stream of equal monthly payments over the life of the mortgage.

What should be of primary concern when you take out a mortgage? Your payments are the heart of the matter—the size of the payments and the time at which you must make them. Given (i) the amount that you borrowed, (ii) the prevailing interest rate, and (iii) the number and frequency of the payments, you can derive the size of the monthly payments. As a general rule, the longer you take to repay your mortgage, the higher the interest rate charged.

There are some intuitively obvious relationships between the prevailing interest rate, the amount that you borrow, and the size of your monthly mortgage payments. For example, your monthly payments will be greater if the repayment period is shortened, if the amount that you borrow is increased, or if interest rates rise.

But mortgages, like bonds, are full of surprises. You may be very surprised by how much your total repayments increase when interest rates rise, especially if you have a long-term mortgage. You may also be surprised to learn how much more than your original loan you have repaid on a long-term mortgage, especially at high interest rates. If your monthly payments are larger than you can handle easily, you may be tempted to extend the repayment period; but under some conditions you may be surprised at how little your payments decrease even if you double the repayment period. If you decide to repay the remainder of your mortgage prematurely, you may be unpleasantly surprised at how much you still owe. Finally, if you find that your mortgage payments are too high you may be pleasantly surprised to find that you can get considerable relief by increasing the later mortgage payments (when presumably, money will be worth less because of inflation) instead of increasing the life of your mortgage.

Which Is the Better Mortgage?

What makes one mortgage more attractive than another? According to some of the folklore you do best to borrow as little as possible, as briefly as possible; that is, one mortgage is better than another if you pay less interest on it. But in fact when you talk about "total interest cost" you are really mixing apples and oranges. Although "total interest paid" is used as a rough measure of the cost of a mortgage it is not really an accurate measure since "equal" payments made at different times are really *unequal* payments. Dollars received tomorrow are worth less than dollars received today; to compare two payments you must consider the effect of interest. A short-term mortgage with lower interest payments may be appropriate for one person but not for another; a five-year mortgage might cost you much less in interest than a 20-year mortgage, for example, but would be inappropriate for a college freshman who planned to go to medical school and whose ability to repay his loans will change radically in 15 years. Jeff Blyskel and Marie Hodge, in an Op-Ed piece in the *New York Times* (9/22/1988), stated that "a golden rule of consumer finance is to borrow money for as short a term as possible. That keeps total interest costs down." But, as we asserted earlier, total interest cost is not critical—what matters most is the rate of interest charged and how much the borrower needs the money and the use he can make of it. Although a borrower should be concerned with the total interest that is paid, it is important to realize that a savings bank borrowing money from a depositor for 50 years at 5 percent interest will repay a great deal of interest for each dollar that it borrows but will prosper nonetheless if it can lend the money elsewhere at 9 percent.

The point is this: when you take out a mortgage you are obtaining the use of someone else's money, and paying interest for it. Whether it is wise to pay this interest depends upon the use you make of this money.

A Formula Relating Length of Mortgage and Interest to Size of Payment

The terms of mortgages vary, so we will fix our attention on one typical mortgage. We will assume that you initially borrow $100,000 and repay it in $12N$ equal, consecutive, monthly payments, making the first payment one month after you take the loan and the last payment N years after the loan. Interest is compounded monthly at a rate which is equivalent to a real annual rate of i percent.

As usual, we are concerned with the relationships between the important variables—the interest rate (i), the number of years that payments are made (N), and the size of the monthly mortgage payments (P). The precise relationship between P, N and i is given by this formula:

$$100,000 = \frac{1 - (1 + i)^{-N}}{(1 + i)^{-1/12} - 1} P$$

To see how the length of the mortgage, the monthly payment, the interest rate, and the amount of money repaid (loan plus interest) are related, it will be helpful to examine the table below. In that table we assume that a $100,000 mortgage was borrowed for a period of from 4 to 40 years at an interest rate that varies from 4 percent to 16 percent. For each of these sets of variables calculate the size of the monthly payment and the total amount repaid during the life of the mortgage.

If you want to test yourself, try to answer these two questions before looking at Table 4.1.

TABLE 4.1

	4%		8%		12%		16%	
Mortgage Length	Monthly Payment	Total Paid	Monthly Payment	Total Paid	Monthly Payment	Total Paid	Monthly Payment	Total Paid
4 years	$2,255	$108,000	$2,428	$117,000	$2,603	$125,000	$2,780	$133,000
8 years	$1,216	$117,000	$1,400	$134,000	$1,592	$153,000	$1,791	$172,000
12 years	$ 872	$126,000	$1,067	$154,000	$1,277	$184,000	$1,497	$216,000
16 years	$ 702	$135,000	$ 909	$174,000	$1,134	$218,000	$1,372	$263,000
20 years	$ 602	$145,000	$ 819	$197,000	$1,059	$254,000	$1,312	$315,000
24 years	$ 537	$155,000	$ 764	$220,000	$1,016	$293,000	$1,281	$369,000
28 years	$ 491	$165,000	$ 728	$245,000	$ 990	$333,000	$1,264	$425,000
32 years	$ 458	$176,000	$ 703	$270,000	$ 975	$374,000	$1,255	$482,000
36 years	$ 433	$187,000	$ 686	$297,000	$ 965	$417,000	$1,250	$540,000
40 years	$ 414	$198,000	$ 674	$324,000	$ 959	$460,000	$1,248	$599,000

(i) If you have a 40-year mortgage at 16 percent interest, how much money will you eventually repay (including the original $100,000 loan)?

(ii) If you have a 40-year mortgage at 16 percent interest and reduce the term to 36 years, how much will your monthly payments decrease?

You probably suspected that the interest you repay on a long-term mortgage is substantial when compared to your original loan, but unless you are familiar with mortgages you may not realize how substantial it is. With a 28-year mortgage at 12 percent interest you eventually repay $333,000 for your $100,000 mortgage so $233,000 of that is interest. In other words, the interest is considerably more than twice the size of the original loan. A 40-year mortgage at 16 percent would cost you $599,000 or $499,000 in interest, which means your interest cost is just under five times the amount that you originally borrowed.

The Sensitivity of Payment Size to Mortgage Length

If the monthly mortgage payments are more than you can afford, you may be tempted to lower them by extending the term of your mortgage. Extending the life of your mortgage will decrease your payments certainly, but probably by less than you hope. If you are wavering between a 28-year and a 32-year mortgage when interest is 12 percent, you find that making four more years of payments only buys about a 1.5 percent reduction in your monthly payment, or about $15. And at 16 percent interest, increasing the term of the mortgage by four years—from 36 to 40 years—decreases your monthly payments by $2 or about 0.16 percent.

The largest part of an early monthly mortgage payment is used to pay the accrued interest; very little of it is used to repay the original loan. If you look at the anatomy of mortgage payments, you will see why this is so and why the monthly payments hardly change when you make a long-term mortgage even longer, especially at higher interest rates. Consider a 28-year mortgage of $100,000 at 12 percent interest with monthly payments of $990. Initially, the monthly interest is about 0.949 percent. This is because $1.00949^{12} = 1.12$. (You might be tempted to use $1/12$ of 12 percent or 1 percent as the monthly equivalent of an annual 12-percent interest rate but that would be incorrect. If the monthly interest were 1 percent a month the annual equivalent would be $(1.01^{12} - 1) = 12.68$ percent a year.)

After a month , 0.949 percent of $100,000, or $949, of your first payment is allotted to paying the interest that has accrued in that first month, and only the remaining $41—a bit more than 4 percent of your total mortgage payment—is left to reduce your debt. Of course in the next month your debt is no longer $100,000 but $(100,000 − 41)$, so the interest accrued in the second month will be a trifle less than that accrued in the first one. In time, your debt will decrease at an accelerating rate but it is very slow going at first. For the last mortgage payments there is a complete reversal; just as the early monthly payments are composed of almost all interest, the later payments are almost all principal.

Length of	Interest Rate			
Mortgage	4%	8%	12%	16%
8 years	26.9%	46.0%	59.6%	69.5%
16 years	46.6%	70.8%	83.7%	90.7%
24 years	61.0%	84.2%	93.4%	97.2%
32 years	71.5%	91.5%	97.3%	99.2%
40 years	79.1%	95.5%	98.9%	99.7%

TABLE 4.2

In Table 4.2 the entries indicate what percentage of your first monthly payment is interest for various interest rates and mortgage terms.

No matter how long the term of the mortgage, the constant monthly mortgage payment must exceed a month's interest on the initial mortgage loan. If it did not, the debt would increase and the mortgage could never be repaid. This means that a $100,000 mortgage at 12 percent interest must have monthly payments that exceed $949—a month's interest on $100,000 at 12 percent. So if the present monthly payment on your 40-year mortgage is $959, you can only reduce your payments by $10 per month no matter how much you extend your mortgage.

When interest rates are high it is obvious that a large part of your monthly payment is devoted to paying interest; the actual percentage may be surprising, however. If there were no interest at all on a four-year mortgage (of 48 payments) your monthly payments would be $100,000/48 = $2,083. Since your monthly payments at 4 percent interest are $2,255, the extra $172 is due to interest. At 8 percent interest on the same mortgage your monthly payment is $2,428 and the extra $345 is due to interest. This seems fair enough; when the interest rate doubles from 4 percent to 8 percent the interest portion of your pension payment just about doubles as well.

For longer-term mortgages, however, the numbers are not as well-behaved. An interest-free 40-year mortgage would have monthly payments of $100,000/480 = $208. A 40-year mortgage at 4 percent has monthly payments of $414, so $206 is interest. If the interest portion of the monthly payment were proportional to the prevailing interest rate, you would expect to pay four times as much interest when interest was at 16 percent rather than at 4 percent; in fact, you pay more than five times as much for interest—that is, $1,040.

The total amount of money you pay when interest rates are high is often underestimated, especially for long-term mortgages. If interest were 0 percent, the monthly payments on a 40-year mortgage would be one-tenth of those on a four-year mortgage since you are paying over a period that is ten times as long. With interest at 4 percent the 40-year monthly payment of $414 is more than 18 percent of the four-year monthly payment of $2,255 (instead of 10 percent of it). At 16 percent interest, the 40-year monthly payment is more than 44 percent of the four-year monthly payment. While it is true that the longer the term of the mortgage the smaller the size of the monthly payments, interest rates affect long-term mortgage payments much more than they do short-term ones. This explains why lengthening the mortgage term decreases your payments much less than you might expect.

Increase Your Payments—Shorten the Length of Your Mortgage

If you find that increasing the term of your mortgage hardly decreases your mortgage payments, you might consider reversing your position by modestly increasing your monthly mortgage payments, thereby cutting the term of your mortgage substantially.

Suppose interest is at 16 percent, and you are considering a 40-year mortgage with monthly payments of $1,248. The monthly interest on $100,000 is about $1,244, so there is no way to reduce your monthly payments by more than $4 per month no matter how long you pay. If you increase your payments from $1,244 to $1,372 (about 10 percent), however, the length of your mortgage will decrease from 40 to 16 years, and the total amount that you must repay will decrease from $599,000 to $263,000—a reduction of more than 50 percent. So the moral is this: if interest rates are

Number of Years until Mortgage Is Repaid	4% Interest	8% Interest	12% Interest	16% Interest
18	$93,100	$95,500	$97,100	$98,100
16	$85,700	$90,200	$93,400	$95,600
14	$77,700	$84,000	$88,700	$92,200
12	$69,100	$76,800	$82,900	$87,700
10	$60,000	$68,300	$75,600	$81,500
8	$49,500	$58,500	$66,500	$73,700
6	$38,600	$47,100	$55,000	$62,100
4	$26,700	$33,700	$40,700	$47,200
2	$13,900	$18,200	$22,600	$27,100

TABLE 4.3

Outstanding Debt Remaining on a $100,000, 20-Year Mortgage

high, you can substantially decrease the length of your mortgage, and the amount that you must repay as well, by raising your monthly payments by only a modest amount.

The Snail's Pace of Debt Repayment

Since the early mortgage payments barely repay the accrued interest, there is little repayment of principal initially. It should not be too surprising, then, that after 15 years of payments on a 30-year mortgage you will not have repaid half of your mortgage loan. But you may be surprised to learn how little you have repaid and how much you still owe. In Tables 4.3 and 4.4 you can find the amount that you owe at various times, and at various interest rates on a 20-year and a 40-year mortgage of $100,000.

Notice that for a 20-year mortgage the amount that is still owed after ten years is considerably more than 50 percent of the original $100,000 mortgage. The amount you owe varies from 60 percent to 81.5 percent, depending on the interest rate. The higher the interest rate, the more slowly your debt is repaid. After two years, which is 10 percent of the term of your original mortgage, you have repaid less than 2 percent of your loan when the

TABLE 4.4				
Number of Years until Mortgage Is Repaid	4% Interest	8% Interest	12% Interest	16% Interest
36	$99,500	$98,300	$99,400	$99,800
32	$90,300	$95,900	$98,400	$99,400
28	$84,200	$92,700	$96,900	$98,700
24	$77,000	$88,300	$94,400	$97,400
20	$68,700	$82,300	$90,000	$95,100
16	$58,900	$74,200	$84,600	$90,900
12	$47,400	$63,200	$75,100	$83,400
8	$34,000	$48,200	$60,300	$69,700
4	$18,300	$27,800	$36,800	$44,900

Outstanding Debt Remaining on a $100,000, 40-Year Mortgage

interest rate is 16 percent. You do much better at 4 percent interest; after two years you will have repaid almost 7 percent of your loan. If you have a 40-year loan, the effect is even more dramatic.

The pace of debt repayment for longer term mortgages is even slower. After 20 years—half the term of the mortgage—the amount you have repaid varies from 5 percent to 32 percent, depending upon the interest rate. With a 40-year, 16-percent mortgage, after four years—10 percent of the life of the mortgage—you have repaid about $1/5$ of 1 percent of your debt, and even after 36 years you still have almost half of your debt to repay.

The Effect of Escalating Mortgage Payments

If the monthly payments on a mortgage are larger than you can afford, there is a temptation to lower them by stretching out the payments. But we have observed that when interest is high or the mortgage is already long-term (or both) the reduction in your payments will be minimal. Fortunately, there is a more effective way to lower these payments.

Most people with long-term mortgages make the same monthly payments throughout the life of their mortgage. Superficially, this seems to

Annual Percentage Increase in Payments	Amount of First Payment
0.0%	$714
0.5%	$680
1.0%	$647
1.5%	$614
2.0%	$582
2.5%	$551
3.0%	$522
3.5%	$493
4.0%	$465
4.5%	$438
5.0%	$412

TABLE 4.5

spread the financial burden equally over time, but in fact the last payment you make on a mortgage (as much as 35 years after you have made your first payment) is usually considerably less painful than your first payment. If, for example, inflation was at a constant 5 percent over 35 years, your first payment would have more than $5\,^1/_2$ times the purchasing power of the last one. If you increase your mortgage payments each year by a small percentage (less than the rate of inflation), the real burden of the mortgage payments would be spread more uniformly and your initial payments would be considerably smaller.

Suppose you take a $100,000 mortgage for N years at a true annual rate of i percent and you make a monthly increase in payment that comes to a true increase of j percent annually. If $R = (1 + j)/(1 + i)$ then the initial monthly payment, P, will be given by this formula:

$$P = \frac{100{,}000\,(R^{-1/12} - 1)}{1 - R^N}$$

In Table 4.5, initial monthly payments are listed for a $100,000, 30-year mortgage at 8 percent interest for various annual mortgage payment increases from 0 percent to 5 percent. An annual increase of 2 percent, for example, means that someone paying $1,000 per month the first year would pay $1,020 per month the second year, $1,040.40 per month the third year, and so on. Given the size of the mortgage, the prevailing interest rate and the length of the mortgage, the size of the first monthly mortgage payment will depend upon the annual rate at which these mortgage payments are increasing.

Notice that by increasing your monthly payment 3 percent every year, your initial payment is reduced from $714 to $522—a reduction of over 25 percent. Your later payments will be greater, of course, but if inflation averages 3 percent or more the financial burden of making these payments will not increase.

Chapter 5

Retirement

{1} Assume that you just retired with $1,000,000 in your pocket and are currently spending $50,000 a year. Maintaining your living standard requires increasing your expenditures each year because of inflation. At the beginning of each year you set aside that year's estimated expenses and invest the rest of your money at 6-percent interest. Estimate how long your money will last if the inflation rate is

    i. 5 percent   **ii.** 4 percent   **iii.** 3 percent   **iv.** 2 percent?

{2} To obtain the capital for your retirement you save $1,000 a year for 45 years. Your capital grows at an annual rate of 10 percent. At retirement you will have contributed $45,000. Estimate the total amount you will have at retirement if interest is added to your contributions.

    i. $100,000   **ii.** $300,000   **iii.** $800,000

{3} Assume you just retired on a fixed pension of $50,000 per year and a lump sum of $200,000. You earn 10 percent on any surplus money that you have and inflation is a constant 4 percent. If you want to live at a fixed standard of living indefinitely (that is, the way you lived during the first year of retirement), about how much can you afford to spend your first year?

    i. $20,000   **ii.** $30,000   **iii.** $40,000   **iv.** $50,000

Answers on page 189.

Home-Made Pensions

Annual income twenty pounds, annual expenditure nineteen nineteen six, result happiness. Annual income twenty pounds, annual expenditure twenty pounds, ought and six, result misery.

—Mr. Micawber in *David Copperfield*, Charles Dickens, 1850

Retirement—A Marathon, Not a Sprint

If Mr. Micawber was right, if the way to avoid misery is to balance your budget, then happiness is easily achieved as long as you are employed. When inflation raises prices you can make do with less, you can work longer hours, or you can get a second job. In any case, periodic pay raises tend to offset, and usually exceed, the erosive effect of inflation. But unless you are one of the lucky few who have their pensions tied to the cost of living, your position is less flexible when you retire. If you have a fixed pension and spend all of it during the early years of retirement, you must expect to see your purchasing power drop as inflation takes hold. The alternative is to pace yourself like a marathon runner, spending less than you earn in the early years of retirement and building up a financial cushion so that you can maintain an approximately constant standard of living.

For most of us the goal is to maintain a level standard of living throughout retirement rather than get by on less and less as the purchasing power of our money dwindles. Despite the apparent precision of the calculations, you should always be aware that pension planning is not an exact science; to plan ahead assumptions must be made about certain variables—inflation and interest, for example—whose future behavior is unknown and unknowable. Even the phrase "maintaining a constant living standard" is virtually meaningless over extended periods of time. Our grandparents did not use computers, nor do we ride in horse-drawn buggies and who is to say which of us has had a better life. In addition, though we may seek to maintain a constant living standard, that may require at times very different incomes. A medical catastrophe, for example, may require a sudden supply of money just to enable you to stay in the same place. To some extent you can avoid the problem by assuming that you pay a constant medical insurance premium and in return the insurance company absorbs all the bumps, but the fact is that in real life there generally will be more than enough bumps to go around even when you are compensated by insurance.

But even with the imprecision it is worthwhile making tentative assumptions about future inflation and interest rates and observing their effect on your pension. By changing assumptions about interest and inflation rates you can observe the robustness of your pension under various conditions. And if in practice your assumptions prove too optimistic, you will be wise to discover this *early* and lower your expenditures accordingly.

Different people retire under different circumstances. Some have accumulated capital, some receive a fixed annuity (fixed periodic payments for as long as they live), and many have a mixture of both. Determining how all these retirees can obtain a stable retirement income (in purchasing power) involves solving a number of mathematical problems. We will focus on three of them described below.

Retirement Strategies

(1) Suppose your sole asset at retirement is a lump sum and your entire retirement income is derived from the interest earned on that lump sum (assuming the interest rate does not change after retirement and is known in advance). If you maintain approximately the same living standard throughout your retirement and there is a fixed, known inflation rate, how long will your pension last?

(2) You are planning in advance to retire with a lump sum as described in (1) above. You know the amount of money that you want to have at retirement—this lump sum will consist of the money you save and the interest that it earns. If you save the same fixed amount each year, what fixed amount should that be to guarantee that you will have your target sum at retirement? If you save a variable amount, paying more in your later years than you do in your earlier ones, and increasing your savings each year by a constant percentage, how will that lighten your burden in the earlier years?

(3) Assume that you have some capital at retirement as well as an annuity, a fixed amount of money that you receive each year as long as you live. What standard of living can you maintain indefinitely assuming that there are fixed inflation and interest rates and you know them both?

Retiring on a Lump-Sum

Let's start simply. Suppose you have saved $1,000,000 at retirement and you currently spend $50,000 a year (as usual we ignore the effect of taxes—things are complicated enough without them). If you want to live as well in the future as you are living now, how long can you survive?

As a first approximation of your pension's endurance you might divide your $1,000,000 capital by your annual budget of $50,000 and conclude that your pension will last 20 years. But that calculation is wrong on two counts. The good news is that the unspent capital—$950,000 in your first year—is earning interest and that will increase your capital. The bad news is that inflation will raise prices over time and force you to spend more money in the future to get the same goods and services that you are buying now. Let's assign some arbitrary numbers to interest and inflation and follow the financial trail for a few years to see how your capital changes.

		TABLE 5.1	

Year	Capital at Start of the Year	Year's Expenses	End of Year Capital (Less Expenses with Interest)
n	$X(n)$	$50,000	$X(n+1) = [X(n) - 50,000] [1.06]$
1	$1,000,000	$50,000	$1,007,000
2	$1,007,000	$50,000	$1,014,420
3	$1,014,420	$50,000	$1,022,285
4	$1,022,285	$50,000	$1,030,622
5	$1,030,622	$50,000	$1,039,460

Five-Year Capital Growth with no Inflation and 6 Percent Interest

Suppose you get 6 percent interest on the money that you invest and the inflation rate is 5 percent. We will assume that at the beginning of each year you put the entire year's expenditures aside and invest the remainder of your capital at 6 percent.

At the beginning of the first year you set aside $50,000 for expenses and receive interest on the remaining $950,000; at the end of the first year you will have $950,000 × 1.06 = $1,007,000.

The following year you put away $50,000 × 1.05 = $52,500 for expenses; this includes a 5-percent increase in the cost of living due to inflation. On the remaining $1,007,000 − 52,500 = $954,500 you get 6 percent interest which comes to 1.06 × ($954,500) = $1,011,770, and this is your capital at the end of your second year. We can continue to calculate your capital in all subsequent years in this way unless and until your money runs out. Table 5.1 summarizes the history for the first five years in the case where there is no inflation.

If you examine your prospects intuitively (when there is inflation) without making any formal calculations, you can take either of two different views. On the down side, your pension definitely is being whittled away by inflation. A 5-percent inflation rate, for example, means that in fourteen years your present fixed pension will only buy about half of what it buys now. On the other hand, looking at the calculations for the first couple of

years, the interest earned seems to overwhelm the erosion caused by inflation; in two years your capital increases by almost $12,000. As it happens,
the pessimistic view is the correct one; despite the financial gains made in
the first two years, inflation eventually wipes out all of your initial capital. To
see why this is so we must construct a financial model.

Your Capital After N Years—A Formula

If you are mathematically inclined, you would first express the relationship
between the capital you have at the end of one year and the capital you have
at the end of the next year by a recursion formula (a recursion formula tells
you the capital you have at the end of any particular year, not as a fixed number, but in terms of the capital you had at the end of the previous year):
$X(n + 1) = [X(n) - 50,000 \, (1.05)^n] \, [1.06]$ where $X(n)$ denotes your capital
at the end of the nth year so that $X(0) = \$1,000,000$. In general, the recur-
sion formula is $X(n + 1) = [X(n) - E \, (1 + j)^n][1 + i]$ where i = the interest
rate on your capital, j = the annual inflation rate, n = the year, $X(n)$ = your
capital at the end of the nth year, $C = X(0)$ = your starting capital, and
E = the amount you spend the first year.

   It is possible to solve this recursion equation to obtain your capital in
the nth year explicitly; the solution turns out to be

$$X(n) = [C - \frac{1 + i}{i - j} \, E] \, [1 + i]^n + \frac{1 + i}{i - j} \, [1 + j]^n$$

   In simple English, you start with C dollars. Each year you buy as much
as you bought the first year but pay more for it because of a j-percent inflation
rate. The money you have earns i percent interest. $X(n)$ is the money you have
at the end of the nth year after retirement. Table 5.2 shows the changes in
your capital when there is 5 percent inflation and interest is 6 percent.

	TABLE 5.2		
Year	Capital at Start of the Year	Year's Expenses	End of Year Capital (Less Expenses with Interest)
n	$X(n)$	$\$50{,}000\,(1.05)^{n-1}$	$X(n+1) = [X(n) - 50{,}000\,(1.05)^{n}]\,[1.06]$
1	$1,000,000	$50,000	$1,007,000
2	$1,007,000	$52,500	$1,011,770
3	$1,011,770	$55,125	$1,014,044
4	$1,014,044	$57,881	$1,013,532
5	$1,013,532	$60,775	$1,009,923

Five-Year Capital Growth with 5 Percent Inflation and 6 Percent Interest

When Will Your Money Run Out?

Perhaps the most pressing question for a pensioner is "When, if ever, will I run out of money?" Stated more mathematically, for what n^* will $X(n^*) = 0$? If we solve for n^* in the equation $X(n^*) = 0$ and call $k = (1 + i)\,/\,(1 + j)$ it turns out that

$$n^* = \frac{-\log\{1 - [C(k-1)]\,/\,[Ek]\}}{\log k}$$

where $\log k$ denotes the logarithm of k.

The entries in Table 5.3 indicate the number of years your pension will last given various inflation and interest rates. It is assumed that you spend 5 percent of your initial capital in the first year and continue to spend enough to keep up with inflation.

If you examine the table carefully you can see certain patterns. One thing you may notice is that if the inflation and interest rates are equal, your pension will last about twenty years. This is because the interest you earn effectively cancels the inflation erosion (but not exactly, since in this model you deduct the expenses at the beginning of the year and collect interest at the end of it). Effectively, you are spending about 5 percent of your pension every year, so your pension will last about twenty years.

The blank spaces in the table represent the cases for which the pension payments will continue indefinitely. If you can afford it, you may want to

	TABLE 5.3									
	Inflation Rate j									
Interest Rate i	0.01	0.02	0.03	0.04	0.05	0.06	0.07	0.08	0.09	0.10
0.01	20.0	18.3	17.0	15.9	15.0	14.2	13.6	13.0	12.4	12.0
0.02	22.1	20.0	18.3	17.0	15.9	15.0	14.3	3.6	13.0	12.5
0.03	25.0	22.1	20.0	18.3	17.0	16.0	15.1	14.3	13.6	13.0
0.04	29.3	25.0	22.1	20.0	18.4	17.1	16.0	15.1	14.3	13.7
0.05	36.8	29.2	24.9	22.2	20.0	18.4	17.1	16.0	15.1	14.3
0.06	58.9	36.4	29.0	24.8	22.0	20.0	18.4	17.1	16.1	15.2
0.07		56.5	36.0	28.9	24.8	22.0	20.0	18.4	17.1	16.1
0.08			54.5	35.7	28.7	24.7	22.0	20.0	18.4	17.2
0.09				52.7	35.3	28.6	24.7	22.0	20.0	18.4
0.10					51.2	35.0	28.5	24.6	22.0	20.0

The Duration of a Pension for Various Inflation and Interest Rates

restrict your spending so that you can support yourself indefinitely. Not only does this alleviate anxiety, but it permits you to pass your capital on to an heir as well.

If $j > i$, that is, if the inflation rate exceeds the interest rate, payments must eventually end no matter how much capital you had originally. This is so because in the explicit formula for $X(n)$ the magnitude of the $(1 + j)^n$ term will eventually be larger than that of the $(1+i)^n$ term, and the coefficient of the $(1 + j)^n$ term is negative. If your payments are to continue indefinitely, not only $i > j$ but the coefficient of the $(1+i)^n$ term must be positive, that is, $C/E > (1 + i) / (i-j)$.

In our example, $C/E = 20$ (C/E is the fraction of your capital you spent in the first year), and the inequality becomes $19i > 1 + 20j$. If $i = 0.09$, for example, then for $j < 0.0355$, the payments would continue forever. This means that if you initially spend 5 percent of your capital and you earn 9 percent interest on your savings, your payments will continue forever if the inflation rate does not exceed about 3.5 percent. If $i = 0.08$ and $j = 0.03$, the inequality is almost satisfied but not quite; you will run out of money but it will take 54 years. In general, to receive payments indefinitely, $i > [1 + jC / E] / [C / E - 1]$. If you only take 4 percent rather than 5 percent of your capital the first year, so that $C/E = 25$ and $i = 9$ percent, then you will still receive your payments forever if j is as much as 4.65 percent.

There is another inference you can draw if you study Table 5.3 carefully. You can estimate fairly accurately when your money will run out without knowing both the interest and inflation rates; it is enough to know their difference, $i - j$. If, for example, the inflation rate is 2 percent less than the interest rate, your pension will last about 25 years.

Although the tables are based upon the assumption that you withdraw 5 percent of your capital the first year, you can easily calculate the duration of your pension payments from the formula, given the inflation and interest rates. The fraction of your capital you withdraw the first year is critical in determining how long your pension payments will last.

Suppose that you get a 10-percent return on your savings. If you only spend 2 percent of your capital in the first year your pension will last forever if inflation does not exceed 7.8 percent. If you withdraw 9 percent of your capital in the first year you will run out of money unless inflation is less than 1 percent. If you withdraw 10 percent or more of your capital initially, any positive inflation rate will cause your payments to end in a finite amount of time.

Neither interest rates nor inflation rates really remain constant for very long, of course. By studying past history, however, you can get some idea of how much these rates vary, and using a mathematical model you can investigate what will happen by assuming various scenarios. Moreover, if what you anticipate deviates from what actually happens, you can modify your behavior accordingly.

Accumulating the Capital for Retirement

If you have no pension at retirement and expect to live solely on the interest earned on a lump sum of money, you will want to know how long your money will last. But at an earlier time you should have been concerned with a related problem—"How do I acquire that lump sum?" You may be sur-

TABLE 5.4

Interest	Number of Annual Dollar Payments n				
Rate i	10	20	30	40	50
0.02	11	25	41	62	86
0.04	13	31	58	99	159
0.06	14	39	84	164	308
0.08	16	49	122	280	620
0.10	18	63	181	487	1,280
0.12	20	81	270	859	2,688
0.14	22	104	407	1,530	5,694

Accumulations at a Constant Savings Rate

prised by how much you can accumulate over a long period of time, especially at high interest rates.

We will investigate two savings schemes and compare the lump sums they accumulate. We will first assume that someone saves $1 at the beginning of every year, for n years, and then calculate the amount that is accumulated. If the savings earn i percent annually for n years the amount accumulated after n years, A, is given by the formula $A = [(1 + i)^{n+1} - (1 + i)] / i$; the values of A are shown in Table 5.4 for various interest rates (i) and periods of accumulation durations (n).

Clearly, the higher the interest rate and the longer you save, the more money you accumulate. But the extent to which a high interest rates affect the amount accumulated, especially over long periods of time, may not be so obvious. Suppose you save your money for 20 years. You will accumulate $31 at 4-percent interest, and $49 at 8-percent interest; since your contribution was $20 in both cases, the interest you earned was $11 and $29, respectively. The point is that although 8 percent is twice 4 percent, the interest earned at 8 percent is substantially more than double the interest earned at 4 percent. If you compare the interest earned at 4 percent and 8 percent over 50 rather than 20 years, the amounts of interest earned is $109 and $570, at 4 and 8 percent, respectively. In this case you earn more than five times as much in interest when you double the interest rate.

For those unfamiliar with the effect of compound interest the results may seem startling, especially over long periods. If you save for 25 years, you accumulate $79 at 8 percent interest; $25 of this is your contribution and the remaining $54 is interest. The interest earned at 8 percent may not seem like much in one year, but after 25 years, more than $2/3$ of your capital is accumulated interest. In 50 years you have $620 at 8 percent interest and the interest earned is more than 90 percent of your capital.

If you want to have some fixed sum at retirement you can link the date of your retirement to the success of your retirement strategy. Suppose, for example, you want to have $1,200,000 when you retire, and save $4,000 a year for that purpose; if you expect to earn 10 percent interest, you should also expect to retire in 35 years. If, contrary to expectations, you only earn 8 percent on your money, you may have to work for 42 years before you retire. But if you are fortunate enough to earn 12 percent, you can retire after 31 years.

Escalating Your Savings

One way to increase the amount of money you have at retirement substantially and painlessly (or to lower your initial contributions) is to increase your contributions over time. It is almost universally true that salaries increase over one's working career and so the effect of raising your pension contribution by, say, 3 percent would be to even out the burden over the years. (When you contribute the same amount of dollars per year throughout your career the actual burden is much greater in your earlier years because of inflation. We justified escalating mortgage payments in Chapter 4 for much the same reason.) Table 5.5 is similar to the one shown earlier in that it shows what you would have after various numbers of years of saving and at various rates of interest. The difference is that you contribute $1 at the beginning of the first year but raise your contribution by j percent each year thereafter. If A is the amount you have after n years, i percent is the prevailing interest rate, and j percent is the increase in your contribution each year, the formula for A is

$$A = [(1 + i / 100)^n - (1 + j / 100)^n] [100 + i] / (i - j)$$

TABLE 5.5

Interest Rate i	Number of Annual Dollar Payments n ($1 the First Year, Increasing by 3% per Year, Subsequently)				
	10	20	30	40	50
0.02	13	33	63	108	173
0.04	14	40	85	160	283
0.06	16	50	117	248	496
0.08	18	62	165	399	918
0.10	20	77	236	660	1,776
0.12	22	98	343	1,117	3,542
0.14	24	124	503	1,924	7,212

Accumulations at an Increasing Savings Rate

In Table 5.5 it is assumed that the contributions are increased by 3 percent every year until retirement, that is, $j = 3$ percent. The entries in the table reflect the amount accumulated for various interest rates and periods of contribution.

Compare the results in Table 5.5 with those of Table 5.4. Increasing your annual dollar contributions by a modest 3 percent a year can lead to a substantial increase in the amount that you accumulate even if the savings period is as short as twenty years. If the interest rate is 2 percent, you would accumulate $33 rather than the $25 you saved without the increase. If you saved for 50 years rather than for 20 years you would accumulate $173 rather than $86.

Retirement on an Annuity and a Lump Sum

Some people retire with only a negligible amount of money but with an annuity, a pension, perhaps, that pays a fixed amount regularly while they are living. Others may have no annuity but a substantial lump sum. And some have both a lump sum and an annuity. But all of them likely share a common concern: Are their retirement incomes sufficient to sustain them indefinitely or will they eventually run out of money?

If a retiree wants to know whether he will be able to maintain a constant living standard indefinitely or if he will eventually run out of money, he can use the same kind of analysis we used earlier whether the source of his retirement income is a lump sum, an annuity, or a mixture of both. The basic idea is that fixed annuity payments are really equivalent to a lump sum.

Suppose that an annuity pays P at the end of every year (we assume the payments continue forever) and the interest rate is, and remains, i percent. The sum of all these payments discounted to the time of retirement is equivalent to the following:

$$P = \left[\frac{1}{(1+i)} + \frac{1}{(1+i)^2} + \frac{1}{(1+i)^3} + \ldots \right] = \frac{P}{i}$$

In effect, an infinite number of constant annuity payments are made equivalent to a lump sum at retirement.

Suppose you retire with an annuity of $50,000/year and with a lump sum of $200,000 and you want to know the most money you can withdraw the first year so that your payments will continue indefinitely to maintain a consistent standard of living. Assume that your expenditures will go up by 4 percent each subsequent year, you earn 10 percent on your money, and the inflation and interest rates will remain unchanged.

The first task is to convert your $50,000/year annuity into an equivalent lump sum. Taking $50,000 / 0.1 we obtain $500,000, so that you are in effect retiring with a $500,000 + $200,000 = $700,000 lump sum. The proportion of this lump sum you can take is given by the formula used earlier: $(i-j) / (1+i) = 0.06 / 1.1 = {}^6/_{110}$. The amount you can withdraw each year, then, would be ${}^6/_{110}$ of $700,000 = $38,182.

If you read this chapter carefully, you might conclude that its foundation is solidly fixed in midair. We usually assumed that inflation and interest rates will be fixed in the future (they won't), we assumed that we know these fixed instant rates (we don't), we assumed that the money required to have a fixed living standard is fairly constant modified only by gradual inflation (we may be traumatized by fire and floods or sudden illness that throw our best–laid plans into a tailspin), etc.

For all of that, you can still draw useful conclusions. If you are saving for retirement, it is helpful to know how much to save each year to reach a retirement goal. If your plans don't materialize—if you are heir to a fortune, if you get less in interest, or if you lose more by inflation than you anticipated—you can always revise your investment plans using new information. You can appreciate that you should increase your savings over time even if you do not know exactly what interest you will earn or how much living costs will rise. The conclusions and the detailed tables are not to be taken literally but are merely suggestive and should be modified by experience.

Chapter 6

The Psychology
of Investing

Test Your Intuition 6

In a short story by Robert Louis Stevenson a magic bottle is described. The imp within that bottle is obliged to fill all the material desires of the bottle's owner—money, jewels, etc. The imp cannot preserve or lengthen the owner's life, however, nor can he change the rules by which the bottle is transferred from one owner to another. The rules are as follows:

The bottle is indestructible: it must be bought for a whole number of dollars (at least one) and for less than the previous owner bought it.

An owner cannot lose the bottle or give it away. If he tries to do either, or sells it for too much, it somehow always returns to his possession.

A buyer must know in advance all the conditions relating to the sale.

If you cannot sell the bottle (and this is the most important condition of all) if you own the bottle and die with it in your possession—you burn in hell forever!

If all goes well, you can buy the bottle, satisfy your material wants, and then sell the bottle at a lower price with nothing lost. (We assume you have a month's grace so that you cannot die within the first month that you take possession.)

Would you be willing to buy the bottle at any price?

Answer on page 189.

The eminent physicist Eugene P. Wigner once observed that scientists were incredibly fortunate in that nature behaves in a manner that can be described by precise mathematical formulas. But anyone who has invested in the stock market knows that the laws of economics are intimately connected to human psychology. It is not clear whether there are laws governing human psychology in the same way that the law of gravity determines the orbits of heavenly bodies, but if they exist they have not yet been found. It is generally accepted that if you want to invest in the stocks of a company, you must not only look into the economic state of that company and predict its future but must anticipate as well how other investors will view that company (some "technical" investors are only concerned with how other investors view that company, as we shall see). This brings us to the subject of this chapter, investor psychology.

The "Proper" Price of a Stock

If you read the financial columns in the newspaper, you undoubtedly have encountered the terms "overvalued" and "undervalued" stocks. These terms imply that a stock has a "proper" price. They also suggest that there is some way to calculate that price. A great deal of effort is expended on such calculations so investors can buy stocks that are below their proper price and sell stocks when they rise above it.

How do you price a stock? Certainly not the way you price a painting. A work of art is evaluated subjectively—its monetary value is determined by what others will pay for it. And opinions about its worth may be widely divergent; a painting greatly valued by its owner may be worth nothing to everyone else. A stock certificate, on the other hand, has little intrinsic value; its worth is derived from the earnings and anticipated earnings of the company it represents.

If a company's future earnings could be known, there would be little difference of opinion about what its stock price should be (there would still be some differences about how to discount future payments, but by and large its value could be calculated). Even if future earnings are not known precisely but are expressed in terms of a probability distribution, there would still be general agreement about the stock's approximate value. Differences of opinion would stem from variations in potential buyers' attitudes toward risk. In fact, we do not know what companies will earn nor can we even describe these future earnings probabilistically with any confidence; we must make do with what we *think* a company will earn—and this is the crux of the problem.

The seeds that determine tomorrow's stock market prices are scattered all about us today, and, given the experience and sophisticated tools available to market professionals, you would think that they could predict the future—and indeed, they do it all the time. It is well-known, for example, that "a strong economy is good for earnings, which is good for stocks." It is also well-known that "a weaker economy lowers interest rates, which is good for stocks." Similar truisms are "acquisitions promise synergy, which is good for stocks" and "divestitures rid companies of losers, usually past acquisitions, which is good for stocks." Finally, "a weakening dollar increases our competitiveness, which is good for stocks" and "a strengthening dollar would encourage foreign investment in the United States which is good for stocks."

These aphorisms from the *New York Times* are offered tongue-in-cheek as examples of a general proposition—*whatever* happens is good for the market. One could just as easily support the opposite proposition—whatever happens is bad for the market. Extreme market pessimists often support their position by maintaining (at different times) that high employment (because it overheats the economy) and low employment (for obvious reasons) are bad for the market.

The point is that the art of predicting future market behavior, or anything else, on the basis of past and current data, is hazardous. After the
atomic bomb was dropped in World War II, newspaper articles described a
new and enormous source of energy that was now available, one that would
make it possible to run the city of Milwaukee on one teaspoonful of seawa-
ter. Thirty years later there were lines of cars surrounding gas stations
because of a gasoline shortage. Some experts stated then that all of the easily
accessible petroleum had already been depleted and the inaccessibility of the
remaining petroleum made it inevitable that prices would remain high in
the future. Some twenty years later gas prices plummeted because of a gasoline glut! And then rose again.

Future Stock Prices—The Fundamental and Technical View

If you believe there is a difference between the "real" value of a stock (the
price it would have if future earnings were known) and its perceived value
(its actual price in the open market), there are two possible views you can
take concerning investing. You can do your own stock evaluations (presumably more insightful than those of most other investors) and buy stocks that
are priced lower than their true market value and wait until the market vin-
dicates your decision.

This is the fundamental approach. But to do this you must somehow
be able to define what the stock is really worth and, when you do, there is no
telling when the market will acknowledge your wisdom and raise your
stock's price to its appropriate level. The other view, the technical approach,
ignores the ideal theoretical value of the stock and focuses on how others,
rightly or wrongly, evaluate the stock. The technician then extrapolates, predicting how these other investors will evaluate the stock in the future based
upon investor's psychological patterns observed in the past. We will say
more about this technical approach presently.

Keynes' Beauty Contest

Some time ago the famous economist John Maynard Keynes observed that investing in the stock market was not just a matter of picking out stocks that are underpriced. To make a profit on a stock you must sell it; there is no point in buying a good stock with good value at a low price if, when you want to sell it, the price is lower, albeit the stock is of even better value. The real question, Keynes suggested, is not what a stock is worth but rather, what people say it is worth, that is, what they are willing to pay. He further suggested that making the right choices when buying a stock is analogous to judging a beauty contest with somewhat unusual rules.

In Keynes' beauty contest there are many judges and each picks a contestant. But it is the judges, not the contestants, who compete. The winning judge is the one whose vote corresponds most closely to the votes of the other judges. In such a case, a judge's opinion of a contestant's beauty is not critical. For the judge to win the contest he must anticipate today how the other judges will vote tomorrow, and vote accordingly; moreover, he should expect that the other judges will be doing exactly the same thing. In short, a judge does not want to know who is most beautiful or even who most judges will think is most beautiful but rather who most judges will think the other judges will select as the most popular selection for most beautiful. A booby prize definition of the technical approach to investing might be described as the art of anticipating now how other investors will behave in the future.

If you accept Keynes' view, as an investor you would do well to recognize and exploit market behavior. But before trying to decipher patterns of stockholder behavior let's first consider whether such patterns really exist. Perhaps short-term market behavior is random, just a multitude of people tossing coins. If there really are such patterns of behavior, how do we find them?

Technical Analysis

One assumption technicians make is that information enters the market gradually and results in trends. As new information is obtained about a stock and gradually spreads, the stock moves from one stable price to another (with minor variations). The process can be detected and analyzed solely on the basis of changing stock prices and the volume of trading—a joint barometer commonly used by technicians—and without any actual knowledge of any new information.

Many different kinds of technical "rules" are offered for prediction of future market behavior on the basis of past market performance. But it is hard to know if the technical approach is justified even when it seems to be successful because these technical rules are widely known and accepted, and often become self-fulfilling prophecies. If, for example, it is generally believed that a certain stock price pattern in the past usually leads to an increase in the stock's price in the future, things may very well turn out just so not necessarily because fundamental new information is available but because advocates of the pattern theory are buying. In fact, much technical analysis is not concerned with whether market action really reflects fundamental new information; it is enough that the buying public is buying. If you can buy a "hot" stock as it starts to rise and sell it somewhere near its high, it doesn't really matter whether the rise was based on fact or fantasy.

Patterns of Human Behavior

It seems pretty clear that there actually are certain patterns that influence an investors' behavior although they are not as precise nor as universally true as Newton's laws of motion.

Common experience and psychological experiments bear this out. Anyone who participates in a weekly poker game knows that it is generally the loser who cannot bear to leave the table and that it is the winner who is first to call it a day. Two psychologists, Daniel Kahneman and Amos Tversky, devised a number of experiments in which subjects acted differently, and consistently so, in the same situation depending only how the situation was described. The following experimental observations confirmed what has long been part of the folklore—people who have been losers tend to become more reckless as they attempt to get even while winners tend to be more conservative and hold on to their profits.

Subjects of an experiment were asked what they would do in each of the following situations:

(1) Imagine your rich aunt died and left you $200. Immediately after that a sporting uncle gives you the following choice—you can collect a certain $50 or have a 25-percent chance of collecting $200.

(2) Your rich aunt dies and leaves you $400. Immediately after that you are apprehended for speeding and the judge says that you can either pay a $150 fine or have a 75-percent chance of paying a $200 fine.

Notice that in both cases you effectively have the same choice: collect $250 with certainty or get $400 with $1/4$ probability and $200 with $3/4$ probability. Yet subjects consistently gambled in case (2) but took the sure $250 in case (1). When they perceived it as a question of gaining (more) money they tended to choose the conservative course; when they formulated the problem in terms of losing money (even though they had just won money), they were willing to gamble to avoid a loss.

The rules that determine how people behave have a rough predictability but are not always rational or consistent with the person's stated goals. Steven Slalop conducted an experiment in which subjects were told to imagine that 600 people were threatened with a disease and that a choice must be made from four possible options (consequences):

(a) Exactly 200 people are saved.

(b) 600 people are saved with $1/3$ probability; none are saved with $2/3$ probability.

(c) Exactly 400 people die.

(d) No one dies with $^1/_3$ probability; everyone dies with $^2/_3$ probability.

Seventy-two percent of the subjects preferred alternative (a) to alternative (b) while only 22 percent preferred alternative (c) to alternative (d) despite the fact that (a) was equivalent to (c), and (b) was equivalent to (d). Slalop then goes on to imagine two litigators pleading their respective cases. The defendant's lawyer claims his client's action was irreproachable since it saved 200 lives with certainty and the plaintiff's lawyer says the same action was reprehensible since it committed 400 people to certain death.

A basic concept of Dow theory, a popular technical notion named after the same Dow who is associated with the Dow-Jones Company, is the resistance level. When a stock sells in a certain price range for a while, then drops substantially, and after some time starts to rise, it often bogs down as it approaches the original price range. The explanation is that investors are reluctant to be losers. The people who bought their stock at the original equilibrium price have been carrying a loss for some time and have regretted their original purchase. When the stock finally returns to the original purchase price these people can hardly wait to sell. According to the theory, if the stock can overcome this resistance, that is, if there are enough buyers to satisfy all the stockholders who want to get out, the stock is said to "break through" a resistance level and is likely to go substantially higher. There may be no fundamental reason why the resistance level should matter; the physical plant, the annual profits, and the future prospects of one company may be as bright as another. But if there are many people who bought the stock when the price was higher (the resistance level), and if they are determined to sell as soon as the stock reaches the price they paid for it, this is of technical importance because the stock cannot rise above the resistance level until they have sold all their stock. If asked, most stockholders would say their main purpose is to make profits and a stock now selling for $50 is not more or less attractive because they bought it at $70 rather than $30. If you are rational, and not ego-involved, your decision whether to hold or sell a stock should not depend upon the price at which you bought it (tax consequences aside). If this theoretical approach to resistance levels is valid, it seems to imply that a substantial number of stockholders are irrational.

Another rule that often governs stockholder behavior, "the law of the maturity of chances," is also irrational. If thirty consecutive even numbers turn up on a roulette wheel it is widely believed that the next number is more likely to be odd. The justification goes something like this—in the long run, if the wheel is unbiased, approximately the same number of even and odd numbers should turn up. Since the even numbers have a "head start" the odd numbers are likely to be impatient and their emergence should be imminent. (A more rational approach would be to question your assumption about the wheel's lack of bias and to conjecture that the next number might be more likely to be even.) But even if the wheel really is unbiased it does not remember what happened earlier—even and odd remain equally likely. Stockholders often indulge in this kind of reasoning when, after seeing their pet stock fall many days in succession, they conclude their stock is "due" to rise since Lady Luck abhors an imbalance in ups and downs.

The important question is not whether stockholders should act this way but whether they actually do act this way. The testimony of the experts is mixed. You can perform your own personal experiment. Imagine that you bought a stock for $50 and, after watching it fall to $30, you see it struggle back to $45. If by then you have become thoroughly disenchanted with the stock, would you sell immediately at $45 or would you be tempted to wait until the stock inched its way back to breakeven at $50 so you wouldn't take even a small loss? Experimental evidence indicates that if you wanted to break even you would have lots of company, as suggested by an article in the *New York Times* of February 21, 1999. Reporter Mark Hulbert cited a study by a professor of finance, Terrance Odean of the University of California at Davis, who analyzed 10,000 histories of stock traders. Odean found that although the Wall Street adage says, "cut your losses and let your profits run," traders actually do the reverse. Only 9.8 percent in the study accepted their losses while 14.8 percent cashed in their profits. The feeling seemed to be that their losing stocks had suffered their fair share of losses and their winning stocks had reaped their fair share of gains, and in both cases they evidently thought worm was about to turn.

Collective Delusions

Clearly, stockholders as decision-makers have their quirks, and these quirks must be taken into account, but that is only part of the problem. Individuals in an active stock market act like members of a crowd and crowds, as Gustave Le Bon noted some time ago, behave according to their own rules.[1] Le Bon observed that the behavior of a crowd is different from the average behaviors of the persons of which it is composed just as the chemical compound formed when an acid and base combine is different from its components. He asserted that the perceptions of crowds are both simple and exaggerated, and members of a crowd are very open to suggestion. To illustrate the point, Le Bon tells of a ship searching for another from which it had been separated during a storm. When the watch indicated that it had sighted a disabled vessel the entire crew "clearly perceived" a raft filled with men giving distress signals. This was only a consequence of mass hysteria, however; as they neared the object crew members saw "masses of men in motion, stretching out their hands, and heard the dull and confused noise of a great number of voices." What they "saw" was a few branches of trees covered with leaves. When the evidence of this was irrefutable the hallucination vanished.

The suggestibility, gullibility, and irrationality of individuals in any market is aggravated when there is a rapid growth of volume in a short time. When a market booms "it is like discovering a gambling casino where the odds have changed suddenly and dramatically in favor of the patrons. Even those who believe the bubble will burst are tempted to get their feet wet and get out before the dam collapses."[2] This has been confirmed repeatedly—in a runaway real estate market in Florida (some of which was inhabited by fishes), in a tulip craze in the early seventeenth century (when the price of a

---

1. Le Bon, G., *The Crowd: A Study of the Popular Mind*, Viking, 1960 (originally 1895).

single tulip approached that of a house), and countless times on various stock exchanges. Charles Mackay has described the South Sea bubble that occurred in the early eighteenth century.[3] The South Sea Company had high connections in and out of government and a number of prospective "profitable" projects. This financial enterprise was described as "a company for carrying on an undertaking of great advantage, but nobody [is] to know what it is." Mackay goes on to say that

> ... the man of genius who essayed this bold and successful inroad upon public credulity, merely stated in his prospectus that the required capital was half a million pounds, in five thousand shares of 100 pounds each, deposit 21 pounds per share. Each subscriber, paying his deposit, would be entitled to 100 pounds per annum per share. How this immense profit was to be obtained he did not condescend to inform them, but promised in a month full particulars would be duly announced, and a call made for the remaining 981 pounds of the subscription. Next morning at nine o'clock, this great man opened an office in Cornhill. Crowds of people beset his door, and when he shut up at three o'clock he found that no less than one thousand shares had been subscribed for, and the deposit paid.

The man of genius immediately left for the continent and was never heard from again.

Cedric P. Cowing in his account of the stock market crash of 1929, cites two journalists who, at different times, reported on this phenomenon in the *Saturday Evening Post*. Cowing reported a Mr. Atwood's comments in January, 1929: "Recent buying, he said, was not based on reasoning but simply on the fact that prices have risen; a rise had led the public to expect more

---

2. Dreman, D., *Psychology and the Stock Market*, Anacom, 1977.

3. Mackay, C., *Extraordinary Popular Delusions and the Madness of Crowds*, L. C. Page & Co., 1932 (originally 1841).

and more returns. The public was blind to think that a 25–30 point rise in a stock in one year could be justified by a 1–5 percent increase in earnings."

Cowing also cited Edwin S. LeFevre (*Saturday Evening Post* 4/9/1932 and 5/2/1936): "LeFevre was inclined to think that the average American preferred to speculate rather than invest; it was part of the American huckster instinct, he said. Neighbors bragging about winnings was a greater stimulus in America than the brokers and tipsters."

It is difficult to keep your head when popular stocks are soaring in a booming market, even (or perhaps especially) if you are a professional investor responsible for the funds of others. In 1972 the Chase Manhattan Bank had been favoring sound stocks and ignoring the fashionable ones but it gave up the search for the value stocks in pursuit of high-multiple growth stocks just in time "to run smack into a swinging door."

Are market prices determined by hysteria, emotion, and the fashion of the moment, or is there method to the madness? Are there enough informed professionals to ensure that stocks take on their "appropriate" prices and to keep the markets efficient? Some economists—Charles Lee, Andrei Shliefer, and Richard Thaler—take the view that there are anomalies in the market that indicate otherwise. To prove their point, they make use of closed-end mutual funds.

A closed-end mutual fund consists of a basket of stocks whose prices are fixed by a competitive market. When you buy a closed-end mutual fund you own a share of that basket and must sell it to someone who, in turn, will buy that share but who also has no direct control of the individual stocks themselves. The price of the mutual funds and the price of the component stocks are fixed independently, each in its own competitive market. It is a routine calculation to find out what the value of each share in the fund "ought to be" using the quantities of each stock in the fund and their prices.

During 1989 these economists found that the costs of mutual funds ranged from a 30-percent discount to a 100-percent premium, but discounting was the rule. The largest such fund, Tricontinental, varied from a

2.5-percent premium to a 25-percent discount. Though there is some distortion introduced by stocks that have capital gains (taxes that are due to previous profits that the purchaser would have to pay when the stocks are sold) the authors assert that the gains are no more than 6 percent of the fund's value and therefore not very significant.

The curious thing is that there seems to be a fashion in premiums/discounts that changes with time. In 1929, closed-end mutual funds sold at a median premium of 47 percent; a premium from 50 percent to 100 percent was considered appropriate but now premiums are much lower and discounts more common.

Suppose it is true (as many say) that investors are gullible and often act irrationally. Suppose it is true that investors are not concerned whether tulips, land values, and stock prices are rising in the long run, that it is enough for them that other buyers think they are rising in the short run. Finally, suppose it is true that many investors act in concert (perhaps unconsciously) and often overreact to both good and bad news. Assuming all of this, how does one exploit this information?

R. L. Stevenson's Genie—A Model for a Manic Market

In "The Bottle Imp," a short story by Robert Louis Stevenson, introduced at the opening of this chapter, a character finds himself in much the same position as someone involved in a feeding frenzy on Wall Street. Let's review the rules of the magic bottle.

A person who owned the bottle of the story had at his service an imp that could satisfy all his material wishes. The imp could give its master all the money and worldly goods that he desired; it could not prolong life, however, nor could it change the rules which determined how the bottle was transferred. The bottle itself was indestructible and the owner of the bottle had to acquire it by purchasing it from the previous owner for less than the previous selling price. An owner could not lose the bottle nor could he give it away. If he tried, or if he sold it for more than he paid for it, it somehow always returned to his possession. An interested potential buyer had to be informed of all the conditions under which he was receiving the bottle or the sale was invalid. And finally—and most important—the downside: if the owner died with the bottle still in his possession, the owner burned in hell forever.

In analyzing this story, we will assume (although the original story did not) that the currency used is the American dollar, that each price must be a positive, integral number of dollars, and at least $1 less than the previous selling price. Thus once the bottle is sold for $1 no future sale is possible. We will also assume (so that we can focus on the real issue) that the buyer will not die immediately and that he will have some opportunity to try to sell the bottle.

Returning to the question posed under *Test Your Intuition 6*, a crucial question is would you consider buying the bottle under any circumstances and, if so, what would you be willing to pay? (If it is helpful, assume the potential buyers are exactly as intelligent as yourself.)

At first it might seem that you have nothing to lose by buying the bottle as long as you pay a sufficiently high price for it. What you paid for the bottle can be recovered immediately by the imp, and once you get whatever you want in addition, you should have no trouble selling the bottle at a slightly lower price to a person who reasons just as you do. But if you look closer, the argument seems less compelling.

It is clearly suicidal to purchase the bottle for $1 since that guarantees catastrophe. Nor are you in a much better position if you purchase the bottle for $2; theoretically it might be possible to sell the bottle at the lower price of $1 but the buyer would have to be informed that this was the last possible transaction and therefore his purchase will necessarily condemn him to eternal damnation. Under such conditions finding a buyer would be difficult.

In fact, the bottle is dangerous to buy at any price. A strategy that prescribes the purchase of the bottle for at least N but not at a lower price would prove to be self-defeating if everyone adopted it. Once you bought the bottle for N (in accordance with your strategy) no one else would buy it from you (also in accordance with your strategy). If there were any safe price at which to buy the bottle, there would have to be a minimum safe price, and that would be self-contradictory. (If anyone bought the bottle from you at the minimum safe price, how would they sell it at a price below the minimum safe price?)

You can reach the same conclusion in a more informal way. If someone buys the bottle, however high the price, someone must eventually be stuck with it; therefore there would have to be someone who thought the price was sufficiently high to justify buying the bottle but who, in the end, turned out to be mistaken.

Participants in a chain letter scheme are in a position similar to the buyer of the magic bottle. A typical recipient of a chain letter is told to send some money to some number of people who preceded him in the chain and to forward the letter to new people with the same request. The recipient is assured that subsequent recipients will more than compensate for his initial payment.

Initially there is a sea of potential untapped candidates, and the promises are fulfilled, but after a while the potential candidates dry up and the prospect of repayment becomes very doubtful. At the end of the day, many contributors will get nothing for their pains.

An investor who judges that the market is overpriced but is tempted—as he observes others speculating enthusiastically and profitably in a rising market—is much like someone considering the purchase of the bottle containing the imp. There is a temptation to get into the rising market quickly and then get out with a profit—all this before the world comes to its senses. Buying a stock you think overpriced because you believe someone out there will buy it at an even higher price is the same dangerous game as buying the imp in the bottle. While there is no doubt that some have gotten in and out of speculative markets profitably, when the public does come to its senses there will be many who will be surprised and hurt. During the bull market of the late 1920s professionals such as Roger Babson and Bernard Baruch perceived the market as overpriced and stayed out. Although they may have lost some potential profits they felt they could not predict when the market would turn. It is probably a good idea to not rush in where experienced and successful investors fear to tread.

A Mathematical Miscellany

{1} Imagine that a detective agency has been hired to protect two banks. The first, bank I, contains $1,000,000 while the other, bank II, contains $100,000,000. The agency is shorthanded and can only guard one of the banks and the bank knows that the potential thieves are aware of this.

Each of the two adversaries—the agency and the thieves—has two strategies. The agency can guard one bank or the other; the thieves can try to rob one bank or the other. If the thieves approach the bank that is guarded, no money changes hands; if the thieves approach the bank that is unguarded, the thieves get, and the bank loses, all the money the bank contains. Neither the thieves nor the agency knows what their adversaries will do in advance. The thieves want to maximize the average amount of money they will get and the agency wants to minimize it. What should the thieves and agency do?

TABLE 7.1

| | THIEVES | |
	Approach Bank I	Approach Bank II
Guard Bank I	$ 0	$100,000,000
Guard Bank II	$1,000,000	$ 0

(AGENCY)

The matrix entries indicate the amount stolen for each possible strategy pair.

{2} You have 10,000 diamonds of various sizes and are interested in selecting the largest (the second largest is of no more value to you than the smallest). You examine the diamonds, one at a time in random order, and must decide to accept or reject a diamond immediately after you inspect it. If you ever reject the largest diamond or accept a diamond that isn't largest, you lose. When you have seen two diamonds you know which is larger but know nothing of the diamonds you have not seen. If you are clever enough to pick the right strategy, you can make the probability of selecting the largest diamond approximately

    i. $1/10,000$ **ii.** $1/1,000$ **iii.** $1/100$ **iv.** $3/8$

Can you guess what an effective strategy might be?

Answers on page 190.

Miscellaneous Economic Applications of Mathematics/Logic

The garden-variety mathematical calculations associated with mortgages, bonds, and stock options usually come under the heading of "mathematics of finance." But there are also less commonplace techniques that use logic and mathematics to solve subtler problems that arise in economic and financial problems. Let's start with a problem that arises in many different contexts, a problem with a surprising and elegant solution.

Selecting the Optimal Member of a Sequence

Suppose you had to pick exactly one of many options, options that are revealed to you sequentially over time. You might, for example, have N different job offers from which you must select one. We assume that the jobs are offered one at a time and you must accept or reject a job irrevocably *at the time it is offered.* You have perfect recall of the jobs offered earlier (and rejected) and you know the total number of jobs to be offered but you know nothing of the jobs yet to come. When two jobs are described you can tell immediately which is better and *your sole purpose is to select the very best job*; we assume that you are indifferent with respect to the second-best job and the worst one. A similar problem arises when you have to hire one of many employees, sign one of many possible commercial contracts, or buy one of many new houses.

An Optimizing Example: The Venture Capitalist's Problem

A venture capitalist is approached by a number of small engineering firms. Each of them is bidding on the same contract and each needs capital to construct a prototype. The venture capitalist interviews each firm in turn and must reject or offer money on the spot since the competing firms need money immediately and, if put off, will go elsewhere. The capitalist can tell which of two firms is better once he sees them but knows nothing about the firms he has not yet seen. Ultimately only one firm will get the contract and that firm will be the best of the lot. If the firm that is financed is not the best, it won't get the contract and the venture capitalist's entire investment will be lost.

To aggravate the problem we will make the unrealistic assumption that there are 10,000 firms to be interviewed and they are interviewed in random order. What strategy should the venture capitalist use to select the best firm? Using this strategy guess what the probability would be of picking the right firm.

Since there are 10,000 potential firms the venture capitalist can choose it might seem that the chance of reaching the best one is pretty slim. In fact, the chance of success is much better than you might guess, provided you use the right strategy, and this would be so even if there were a million or a billion firms from which to choose.

Before describing the best strategy let's consider a simpler, but inferior, one. We can show that even this inferior strategy is good enough to pick out the best firm *more than a quarter of the time whatever the total number of options.*

To show this strategy is effective, first separate all the firms into two equal groups: group I consisting of that half of the firms you see first and group II consisting of that half of the firms you see last. You automatically reject all the firms in group I, and then accept the first firm in group II that is superior to all the firms that you have examined earlier. (If no such firm appears, you have failed.)

With this strategy you will pick the best firm when the best firm is in group II and the second-best firm is in group I. This will occur a bit more than a quarter of the time. If the third best firm is in group I and the best and second-best firms are in group II (this will occur about $1/8$ of the time) you will still have an even chance of picking the best firm, so you even have extra chances, as well.

Remember, this is not your best strategy. You can do better yet. If you follow a similar strategy but put only about 37 percent of the firms in group I and the rest in group II, your probability of selecting the best firm will rise to about 37 percent; and again, *this is true no matter how many firms you had to select from originally.*

Game Theory

Another weapon in the arsenal of the economic problem-solver is game theory. This is a relatively new branch of mathematics, originally designed as a tool for solving certain kinds of problems in economics. It has since been applied to problems in many other disciplines as well. Oskar Morgenstern and John von Neumann, a distinguished economist and mathematician, respectively, felt that the mathematics generally used to solve economic problems—mathematics generally used for applications in physics—was inappropriate. When you calculate the amount of insulation you will need to keep a house warm in winter you need not worry about a malicious Mother Nature trying to frustrate your plans. But a grocer that lowers his price to increase his sales can be fairly certain his competitors will lower theirs as well.

Game theory addresses the problem of a decision-maker who must make a decision in conjunction with other decision-makers. Each decision-maker has his own axe to grind and each knows that the final outcome will depend upon the choices everyone makes. Game theory attempts to determine what a rational person should do and to predict what the final outcome should be (not always successfully). The best way to learn what game theory is all about is to see how it is used in some simple examples.

Two Business Competitors

You are the representative of a business that is hoping to obtain a contract to manufacture a product and are preparing a model to demonstrate it. Unfortunately, you have a competitor. Both you and your competitor are

---
TABLE 7.2

		YOUR COMPETITOR	
		Add Feature	Reject Feature
YOU	Add Feature	1.0	0.6
	Reject Feature	0.4	1.0

contemplating adding an additional feature (the same feature) to the product but neither of you knows in advance what the other will do.

You and your competitor both believe that your product *as it stands* is better than that of its competition and that you will get the contract if you both add the feature or both omit it. But if only one of you adds the feature, that party, the "adder," will get the contract if and only if the feature is useful. The feature has a probability of 0.6 of being successful.

What should you and your competitor do if you want to maximize, and he wants to minimize, your probability of getting the contract?

In Table 7.2 your strategy consists of picking a horizontal row (the upper row means you add the feature, the lower row means you reject it) and your competitor's strategy consists of picking a vertical column. The choices are made by each of you, simultaneously; each of you makes a decision without knowing what the other did. Your two decisions together determine the probability of your getting the contract. The entries in the matrix represent the probabilities that you will obtain the contract (which you subtract from one to get the probability that your competitor will get the contract).

Suppose, for example, you reject the feature (the lower row) and your competitor adds it (the left column) your chance of getting the contract is 0.4 (the number in the matrix where the second row and the first column intersect).

At first it might seem that you can do no better than to add the feature and (anticipating this) your competitor does his best to reject it. If both of you follow that course, you will get the contract if, and only if, the innovation is a success, that is, 60 percent of the time. If you dig a bit deeper, however, you may notice that there is a chance of doing better. If you feel confident that your competitor will reject the innovation, shouldn't you do the same? If you both reject, you will certainly get the contract whether the innovation is a success or not. The trouble is, of course, your opponent may think deeply, as well, going this far and a bit further. He may anticipate your anticipation and add the feature, in which case you have converted your probability of getting the contract from 0.6 to 0.4. Does this mean you should simply adopt the innovation, as you thought in the first place? Not at all!

Using Mixed Strategies

As we posed the problem originally it appeared that both parties had only two alternatives: to accept the innovation or to reject it. But one of the insights of game theory is that you really have an infinite number of choices: you can add the innovation with probability p, where p is any number between zero and one. Once you choose p, a random device "decides" whether you should innovate or not. If you don't see why this is to your advantage, you will soon. So let's suppose you go beyond the accept/reject decision and instead use a random device to innovate with probability 0.6 and reject with probability 0.4. (We will say more about how we hit upon the probabilities 0.6 and 0.4 later.)

Suppose your competitor adopts the feature—60 percent of the time. You will adopt also and get the contract with certainty; 40 percent of the time you will reject the feature and you will get the contract with probability 0.4. Your overall probability of getting the contract if your competitor adds the feature is $(0.6)(1) + (0.4)(0.4) = 0.76$.

Now suppose your competitor rejects the feature; 60 percent of the time you will add the feature and 60 percent of those times you will get the contract. The other 40 percent of the time you will also reject the feature and you will get the contract with certainty. Your overall probability of getting the contract if your competitor rejects the feature is $(0.6)(0.6) + (0.4)(1) = 0.76$.

So, if you choose the suggested strategy of adopting the feature with probability 0.6 you will get the contract 76 percent of the time *whatever* your competitor does which is certainly better than settling for 60 percent. Moreover, you need not worry about your competitor outwitting you; you can announce your mixed probabilistic strategy in advance and it will not change the odds at all.

Can you do better yet? Afraid not. If your competitor announces in advance that he will add the feature 40 percent of the time you can infer using identical logic that it will not matter whether you adopt or reject. The outcome will be the same: you will get the contract 76 percent of the time. Since either you or your competitor can guarantee unilaterally that you will get the contract 76 percent of the time (on average) this should be the final outcome.

Finding the Right Mixed Strategy

How did we get the idea of using this 60-percent–40-percent strategy? A technique for calculating this strategy (called the minimax strategy) that often (but not always) works, is this. Find the mixed strategy that makes the average outcome the same whatever your adversary does.

In this example, suppose you add the feature with probability p and reject with probability $(1 - p)$. If your competitor adds the innovation, you will get the contract with probability $(p) (1) + (1 - p) (0.4) = 0.4 + 0.6p$. If your competitor rejects the innovation, you will get the contract with probability $(p) (0.6) + (1 - p) (1) = 1 - 0.4p$. Setting $0.4 + 0.6p = 1 - 0.4p$ we get $p = 0.6$. Your competitor's minimax strategy can be derived in the same way.

The Detectives vs. the Bank Robbers

The same game theory methods may be applied in very different situations. Let's return now to the problem we posed at the beginning of this chapter. Imagine that a detective agency has been hired to protect two banks. The first bank, bank I, contains $1 million while the other bank, bank II, contains a $100 million. The agency is shorthanded and can only guard one of the banks and the bank knows that the potential thieves are aware of this.

Each of the two adversaries, the agency and the thieves, has two strategies. The agency can guard one bank or the other; the thieves can try to rob

TABLE 7.3

| | THIEVES | |
	Approach Bank I	Approach Bank II
Guard Bank I	$ 0	$100,000,000
Guard Bank II	$1,000,000	$ 0

one bank or the other. If the thieves approach the bank that is guarded, no money changes hands; if the thieves approach the bank that is unguarded the thieves get, and the bank loses, all the money the bank contains. Neither the thieves nor the agency know what their adversaries will do in advance. The thieves want to maximize the average amount of money they will get, and the agency wants to minimize that amount. What should the thieves and agency do?

In Table 7.3, the two strategies for the agencies are represented by horizontal rows and the two strategies for the thieves are represented by vertical columns. The entries in the matrix represent the amount of money that is stolen for each pair of the agency's and thieves' strategies.

The minimax strategy for the agency is not too surprising. Bank II, which contains most of the money, is guarded most of the time—with probability $100/101$. The thieves' strategy might be considered surprising, however. Willie Sutton, the notorious bank robber, said he robbed banks because that is where the money is. Willie Sutton notwithstanding, the minimax strategy prescribes that the thieves should almost always go to where most of the money isn't. Specifically, the thieves should go to bank I with probability $100/101$ and to bank II with probability of only $1/101$.

The strategies are calculated in the same way as in the previous example. If the minimax strategy is adopted by *either* the agency *or* the thieves (they needn't both adopt minimax strategies, it is enough that just one of them does), the outcome will be an average loss of $\$100,000,000/101$ or just under $1,000,000 to the thieves.

TABLE 7.4

		YOUR COMPETITOR					
		Money Spent on Company A's Prototype in Millions					
		0.0	1.0	2.0	3.0	4.0	5.0
Money Spent on	0.0	0.5	0.0	0.5	1.0	1.0	1.0
Company A's	1.0	1.0	0.5	0.0	0.5	1.0	1.0
Prototype	2.0	1.0	1.0	0.5	0.0	0.5	1.0
in Millions	3.0	1.0	1.0	1.0	0.5	0.0	0.5

(YOU is labeled vertically on the left.)

Optimal Capital Allocation

In my final application of game theory a contract is to be awarded by each of two companies, A and B, and you and a competitor are the only ones interested in either of them. A prototype must be constructed by you and your competitor for each contract. You have $3,000,000 from which you must construct your two prototypes and your competitor has $5,000,000 for building his two prototypes. For simplicity, we will assume each of you must spend some multiple of a million dollars on each prototype. If you and your competitor spend the same amount of money on a prototype, you each receive half of that contract. If one of you spends more on a prototype than the other, the big spender receives the entire contract. How much should each of you spend on each contract and what should the final outcome be (on average) if both you and your competitor want to maximize the number of contracts obtained?

The matrix entries in Table 7.4 represent the number of contracts that you receive. If, for example, you spend $1,000,000 on A's prototype and your competitor spends $2,000,000 on A's prototype, your competitor will win A's contract. Also, you will have $2,000,000 left for B's prototype and your competitor will have $3,000,000 for B's prototype so your competitor will win B's contract as well. If this pair of strategies is chosen, you will receive a total of no contracts which is what the matrix reflects.

To solve this problem you must first apply a little common sense. It is clear that your competitor should never spend more than $4,000,000 on a prototype since that is enough to guarantee the contract. So from an apparent six strategies for your competitor it turns out only four are viable. By symmetry, you can infer that in the minimax strategy the probability of your spending everything on A's prototype is the same as the probability of spending everything on B's prototype, that is, your first and fourth rows have the same probabilities. Also by symmetry your second and third rows have the same probabilities as well. By similar reasoning you can infer that your competitor will play his second column as often as his fifth column and that he will play his third column as often as his fourth column. As mentioned earlier, he should never play his first or sixth column.

Given these clues you can work out the minimax strategies by simple algebra and we leave the calculations to you. It turns out that your competitor should spend $1,000,000 and $4,000,000 on A, each $1/3$ of the time and $2,000,000 and $3,000,000 on A each $1/6$ of the time. You should spend nothing and $3,000,000 on A, each $1/3$ of the time and $1,000,000 and $2,000,000 on A each $1/6$ of the time. On average you will get $7/12$ of a contract and your competitor will get $1^5/12$ of a contract. Notice that if you play your minimax strategy and your competitor spends either nothing or $5,000,000 on A he will only get an average of $1^1/6$ contracts, which is less than the $1^5/12$ contracts which he could have obtained.

Auctions

Another rewarding target for mathematical analysis is the auction. Auctions have various structures and each has its own personality. If you want to sell an object that several potential buyers desire you might consider having an auction. In a silent auction everyone submits a bid at the same time and the person making the highest bid receives the auctioned object and pays the amount that was bid. Some stores vary this theme—they put an object on sale and lower the price each day until the object is purchased. And often bidders are allowed to hear each others bids and respond if they wish by raising their own earlier bids.

Since bidders bid once in a silent auction and then can no longer compete, you might think the seller would prefer a noisy auction. But in noisy auctions the bidders keep their initial bids low so they can compete later. If, in a noisy auction, there is only one bidder who prizes the auctioned object, he will gain his end with a low bid when he observes the lack of competition. This same bidder may make a higher bid in a silent auction since he may be unaware that he is the only interested party.

A seller conducting an auction is essentially passive but presumably will try to choose the kind of an auction that will maximize profits. The buyer in an auction is in a kind of dilemma—too low a bid is unlikely to obtain the desired object. Too high a bid may win the auction but the bidder may be sorry. In general, it seems clear that a potential buyer should bid less than the object is actually worth to him. If he obtains an object at auction by bidding what he thinks the object is worth he has gained nothing. By lowering his bids he lowers his chance of winning the auction but at least he profits when he does win.

Getting Bidders to Bid Full Value

The late Nobel Laureate economist William Vickrey introduced a slight change in auction rules that induced bidders to bid exactly what they think an object is worth—no more and no less. The auction that he modified was the *silent* auction in which everyone independently submits a bid and the auctioned object is exchanged for the highest amount bid. (There is a price the seller pays for using Vickrey's scheme, however; the seller obtains a bit less than the maximum value assigned to the object by the bidders.)[1] Can you figure out a modification of the auction rules that would achieve that end?

Vickrey suggested that the auctioned object be awarded to the highest bidder as it usually is but the price paid for the object by the highest bidder

---

1. In a noisy auction the seller can only hope to receive what the auctioned object is worth to the second highest bidder since the high bidder will have no further competition when that bid is exceeded.

should not be his own highest bid but the *second* highest bid. Why should this rule change make such a difference? It turns out that whatever the other bidders do, a bidder sometimes does better, and never does worse, by bidding what he thinks the object is worth.

To see why this is so, imagine that you feel an object is worth $1,000 but you only bid $950 for it. The outcome will depend upon B—the value of the highest bid made by the other bidders.

If B is less than $950 there will be no harm done; you will get the object and pay B. This is what would have happened if you had bid $1,000.

If B is more than $1,000 there will also be no harm done; you will not get the object but you wouldn't have gotten it even if you had bid its true value (to you), $1,000.

If B is between $950 and $1,000, however, you do not get the object by bidding $950 but you would have obtained it at a value of B (which is less than the $1,000 the object is worth) if you had bid $1,000.

By a similar argument you can show that you never do better bidding more than what you think an object is worth. The outcome will either be the same as it would be if you bid the object's true value or you will pay more for the object than its value.

An Auction That Simulates a Nuclear Arms Race

One last diabolical example of an auction contrived by Martin Shubik, an economist at Yale, takes on some of the overtones of an arms race. A dollar is auctioned off in a customary noisy auction but with one variation on the auction rules: the person who makes the final highest bid keeps the dollar and pays the amount that he bid last, (as usual); but in addition, *the second highest bidder pays the amount that he last bid, as well, and gets nothing!*

It generally turns out that the "winning" bidder often bids more than the dollar he receives and with a little thought you can see why. Suppose that you have made the last bid of $0.95 and have heard that bid raised to $1.00 by a competing bidder. Would you settle for a loss of $0.95 with no compensation or would you re-bid $1.05 hoping for a loss of only $0.05 (on the chance the bidding would end there since your competitor at this point might be as anxious as you end the madness).

The hope of a loss of $0.05 looks very attractive compared to the alternative certain loss of $0.95 that you would have to accept if you decide to stop bidding. Unfortunately the same considerations might induce your competitor to raise his bid to $1.10, thus avoiding the sure loss of a dollar in return for the hope of a loss of only a dime (if *you* drop out then). And the catastrophic logic goes on and on.

The analogy to an arms race is clear. When one of two countries engaged in an arms race decides to improve its military position by investing additional resources to enlarge its arsenal, its adversary, predictably, will do the same. As a result, the relative strength of the adversaries remains unchanged but both of their budgets are strained.

The Winner's Curse

One common characteristic of auctions is that the winning bid often exceeds the value of the auctioned article. This should not come as a surprise. Although people should bid less than what they think an object is worth, at the end of the day the maximum bid is all that matters, and quite often someone is too optimistic. This tendency of the maximum bid to be too high is called the "Winner's Curse." In an article by A. Thaler[2] a jar full of coins was repeatedly offered to classes for auction. There were $8.00 worth of coins in the jar and the average class bid was only $5.13. The average winning bid, however, was $10.01 which caused a loss of $2.01. In the article where the term "Winner's Curse" was first coined[3] it was noted that when bidding for oil fields the greater the number of bidders, the higher the bids tended to be. But the authors of the article felt this was the reverse of how the bidders should behave. If you are a bidder

---

2. Thaler, A., "The Winner's Curse," *Journal of Economic Perspectives*, vol. 2-1, Winter–Spring 1988.

3. Capen, E. C., Clapp, R. V., Cambell, W. M., "Winner's Curse," *Journal of Petroleum Technology*, vol. 23, June 1971, pps. 641–653.

for drilling rights or a publisher trying to acquire a book you might do better to lower your bids as the number of competitors increase. If you made the highest of 15 bids in a field of informed and experienced competitors, wouldn't you have second thoughts about the prudence of your bid if yours was the highest one?

A Future's Market Paradox

We leave our miscellaneous mathematical applications with a paradox about futures which we will simply state and allow you to resolve. Assume that at any particular time the currency of one country can be converted into that of another, but the rate of exchange varies over time. If you can predict the future, if you are clever enough to suspect that your currency will become weaker than some other currency six months from now, for example, you can contract to buy the other currency six months from now at the current rate and make a profit; and therein hangs our tale.

Suppose that a dollar has the same value as a franc today and you know (somehow) that in six months a dollar will either be worth two francs or half a franc; you also know that each of these two possibilities are equally likely. You enter into a futures contract with a Frenchman to buy one French franc for one dollar in six months. Clearly, you and the Frenchman are involved in a zero-sum game. Whatever advantage this agreement yields to you in six months is reflected in a corresponding disadvantage to the Frenchman. Whatever penalty you incur in six months as a result of this contract is reflected in a corresponding advantage to the Frenchman.

If in six months the dollar is worth two francs, you spend a dollar, receive a franc (in accordance with your futures contract) and buy back half a dollar, for a profit of −$0.50 (that is, a loss). If in six months the dollar is worth half a franc you spend a dollar, receive a franc and buy back two dollars, for a profit of $1.00. Since these outcomes are equally likely your expected profit is $(^1/_2)$ (–\$0.50) + $(^1/_2)$ (\$1.00) = \$0.25. So on average you can expect a profit of \$0.25 on your futures contract.

The paradox is this—the Frenchman's position is symmetrical to your own and he too can expect an average profit of $^1/_4$ of a franc. But his interests are diametrically opposed to yours. What he loses, you gain; what he gains, you lose. *You can't both be doing something right.*

This chapter contains a sample of mathematical and logical tools that can be used in a wide variety of economic situations. The technique of choosing the best of a sequence of options might have been of value to General Erwin Rommel in World War II as he allocated his reserves to various sectors as they became suspected targets for invasion. It might also be used by a police inspector allocating his investigative resources to various suspects as evidence turns up in a case. And it might be used by a venture capitalist as we observed. We have seen how game theory can be used by a business competitor, a police department, or a bank robber to choose an optimal strategy and anticipate the future. And we saw how auction rules may be written so that bidders are motivated to make honest bids or to act like participants in an arms race. While hardly exhaustive, these examples give some idea of the utility of these mathematical tools.

Chapter 8

Statistics

Test Your Intuition 8

{1} As an eager investor you are seeking a "miracle stock"—a stock that doubles in three years. From past history you have learned two things: only one stock in a thousand is a miracle stock and your broker is almost remarkably prescient in ferreting out miracle stocks. When shown a miracle stock he will identify it as such 90 percent of the time and if shown a non-miracle stock he will identify it as such 90 percent of the time, as well.

You select a stock at random. If your broker identifies it as a miracle stock what is the probability that it actually is a miracle stock?

    i. 0.9   **ii.** 0.5   **iii.** less than 1 percent

{2} One of your friends claims to be a poker expert. You have access to the daily records at his club and at the end of every day you check whether his *total* poker-playing record (from the first day he started playing poker to the present) shows a profit or a loss.

You find that at the end of 4,000 playing days only 100 days (or 2.5 percent of the total) reflect a losing record and the rest show a winning record (we repeat—not for that one day but for his entire career). What is the probability that he would achieve this record, or an even better record, if he were no better than the others and his performance was due simply to chance?

    i. 0.2   **ii.** 0.1   **iii.** 0.05   **iv.** 0.000002

Answers on page 190.

But the judge said he had never summed up before;
so the Snark undertook it instead,
and summed it so well that it came to far more
than the witnesses ever had said!

—*"The Hunting of the Snark,"* Lewis Carroll

The Economist's Favorite Sport—Jumping to Conclusions

If you read the financial pages of your newspaper, you quickly become aware of the importance of inferential reasoning. Statistics is a critical tool and serves two important functions: it is used to summarize a vast amount of known data by a few key numbers and thus make the data manageable (*descriptive statistics*), and it is used to draw inferences from these numbers (*inferential statistics*). Of course there are also other, less formal, ways of drawing inferences about the world, and we will investigate some of these as well.

Descriptive Statistics—Reducing Many Numbers to a Few

Suppose you are paid a fixed salary and, after long years of service, you are given the option of receiving a fraction of your company's future profits annually instead of your fixed salary. You know the history of your company's profits and you must guess which type of compensation will yield a higher return.

The difficulty is that the profits keep changing. In some years you do better with the fixed salary, in other years you do better sharing profits. What you would really like to do is to take all the salaries earned over many years and summarize them with a single number, so that the two methods of compensation options can be compared.

The process of condensing an ocean of data into a few representative numbers is called descriptive statistics. The down side of using a few numbers to represent many is obvious; you lose information. The up side is that the few representative numbers are comprehensible while the original ocean of data is not. And this is so whether you are deciding if people live longer in Virginia or Vermont, or if Audis last longer than Volvos.

Choosing a Best Single Number

Let's start with the simplest case: describing one facet of a population with a single number. If you represent a population with a single number it will generally be a "typical" number, *a measure of central tendency*. There is, however, more than one such central number from which you can reasonably choose. (There are other numbers that may be used to characterize a population such as measures of dispersion, which we will discuss later.)

This single, representative number that you choose should reflect your purpose. If you are installing a telephone system your main concern will likely be that the system not be overloaded; if you know the number of calls made on each of the last 1,000 workdays you might be interested in the *maximum* number of calls occurring in any one day. Similarly, if you are building a bridge, you might use the *maximum* amount of traffic at any one time as your measure if you are concerned that the bridge may collapse or that the traffic may be congested. If you are overseeing a reservoir and are concerned with the water supply or are responsible for funding an amateur theater group, it is the *minimum* water level and the *minimum* checking account balance, respectively, that will be of primary interest. But, for most purposes, to represent a set of numbers you would use some number in the middle of the range—a measure of central tendency.

Measures of Central Tendency

The most common, single, descriptive statistic is undoubtedly the *arithmetic mean* or *average*. If you received 52 weeks of salary you might ask, "What single salary, received every week for 52 weeks would yield the same amount of money that I actually received during the year?" The answer—the sum of all the weekly salaries received divided by 52 would be the arithmetic mean.

But suppose you deposit money in a bank and leave it there to accumulate interest. If the bank pays a different interest rate every year, you get a different answer if you ask a similar question. Each year the bank multiplies the money you had at the start of the year by the interest rate factor $(1 + i/100)$ if they are paying i percent annual interest. What single interest rate factor, received in each of two years, would have the same effect as two different interest rates k and j? This answer here would not be the average, it would be the *geometric mean*—the square root of the product of the two interest rate factors, that is, $(1 + i)^2 = (1 + j)(1 + k)$.

Other measures of central tendency that are sometimes used are the *median* (the middle number of a population) and the *mode* (the most common number in a population). If a town is located along a single road and each house is on some numbered avenue (the number of the avenue reflecting how far north the house is) a fire station should be constructed at the avenue which is the median avenue of all the house locations if you want to minimize the fire trucks traveling distance.[1]

If you wanted to buy a newspaper stand at one train station you would want the mode—the station that most people use. You may be at the north side of town with less population than the south side but this is of no concern to you since you only sell to people who are actually at your station. If you have a shoe store and are building up your inventory, you would be concerned with the mode, the most popular size, as well. If half the people in

---

1. If you wanted to minimize the distance squared, you would put the fire station at the arithmetic mean.

town had big feet and the rest had small feet you wouldn't make many sales if you concentrated on average-sized feet.

Although the arithmetic mean is used most frequently it is often misleading. Suppose you sell cars that cost $20,000 and you expect your typical customer to be earning $100,000 a year. If one person in town earns $500,000,000 a year but each of the remaining 5,000 people in town earn only $20,000 a year you are not likely to make many sales even though the mean salary is almost $120,000 a year. Although it may be misused, when handled with care the average, or arithmetic mean, can be very useful and is also very convenient for mathematical manipulation.

The Robustness of a Statistic
One thing to consider when choosing a statistic is its robustness, the statistic's tendency to change very little if the data on which it is based only changes a little. If 101, 100, and 100 people wear size 6, 8, and 10, respectively, and a single person changes from size 6 to size 10, the mode size jumps from 6 to 10 as well; the average size, in contrast, is much more stable. It goes from 7.99 to 8.01.

Another consideration in choosing a statistic is the effect of outliers, very rare, atypical data. The half-billionaire in the above example will profoundly change some statistics and may have no effect on others and you must decide what effect it should have for your purposes. If you want to predict how an election will go and you think a voter's pocketbook determines his or her vote, you do best to use the median and forget the millionaire. On the other hand, if you want to estimate what people will pay in taxes, a single individual may be the heart of the matter.

The statistic you choose to reflect some aspect of a population may have great practical importance. In 1961, the Bureau of the Census stated that the average U.S. family income was $5,737 while the Office of Business Economics said it was $7,900; it turned out one of them used the median and the other the arithmetic mean.

Measures of Diversity

Once you pick a measure of central tendency you may want to address the question of variation. Although you may consider a temperature of 70 degrees ideal you would probably be unhappy in a country that was 130 degrees half the time and 10 degrees the rest of the time.

Once again, there are different measures of variation. One is the range: the maximum value minus the minimum value. You can also define variation as the average distance from the arithmetic mean to each member of the population. But the most mathematically convenient and most commonly used measure is called the *variance:* the average squared distance between the arithmetic mean and members of the population. Algebraically, if there are N numbers in a population and X_i is the ith number and M is the arithmetic mean then the variance is defined to be

$$Variance = \frac{(X_1 - M)^2 + (X_2 - M)^2 + \ldots + (X_N - M)^2}{N}$$

The standard deviation, the square root of the variance, is also often used because it has the same units as the members of the population.

For example, suppose four mailroom clerks earn 30, 40, 45 and 65 measured in thousands of dollars per year. Their average salary is $(30 + 40 + 45 + 65) / 4 = 45$ thousand dollars per year. The variance would be $[(30 - 45)^2 + (40 - 45)^2 + (45 - 45)^2 + (65 - 45)^2] / 4 = 162.5$ and the standard deviation would be the square root of $162.5 = 12.75$. If their salaries were 42, 45, 46 and 47 their average salary would be the same, 45, but the variance would be $[(42 - 45)^2 + (45 - 45)^2 + (46 - 45)^2 + (47 - 45)^2] / 4 = 3.5$ and the standard deviation would be 1.87. The smaller standard deviation in the second case reflects the smaller diversity in salary.

Inferential Statistics

In April 1927, at Western Electric's Hawthorne Works near Chicago, industrial psychologists began a study that was to continue throughout the early 1930s. Its purpose was to discover how a worker's environment affected his productivity. During the experiment many changes were made in the workplace—the rest break schedule was changed, the lighting was varied, social groups formed by workers were observed and their members interviewed, shorter work days and work weeks were introduced, soup and coffee were served at the morning breaks—and worker productivity was observed before and after each of these changes. With some exceptions, increased productivity seemed to follow each of these changes, and it appeared that a veritable gold mine of techniques had been discovered for improving production.

But there was something dubious about this amazing sequence of improvements. While production increased as each innovation was introduced, the biggest surprise was to come at the end of the experiment, when the workplace reverted to its original, unimproved state. At that time production improved once again! After much head scratching it was concluded that the improved productivity was really due to the increased attention that the workers had received; the particular changes in working conditions were hardly relevant. This classical experiment is well known to industrial psychologists and is known as the *Hawthorne Effect*.

This story has a moral: predicting the future on the basis of past experience is a hazardous business. You can never be sure that you have taken account of all the variables, nor can you be certain about which is cause and which is effect.

Prediction in an Uncertain World

When mathematics is used to describe the way the world works, a probabilistic rather than a deterministic model is often used. Perhaps the probabilistic model reflects some essential reality, that we can only partially explain things. Perhaps it is simply an indication that we do not know enough to construct a deterministic model. At one time it was thought that the physical world was deterministic and that sufficiently accurate measurements in conjunction with Newton's laws would be enough to predict the future as accurately as one wished. Now physicists believe there is an uncertainty built into the essential nature of the Universe and a limit to the accuracy of predictions. In any case, probabilistic assertions are often made: "70 percent of new businesses fail," "1 percent of all stocks double in a year" and "there is a 20-percent chance of rain." The process of deducing the probabilistic world around us from data we observe is called inferential statistics.

The Perils of Making Statistical Inferences

They say that ignorance isn't the enemy; the real enemy is the things you know that aren't so. By reasoning incorrectly you can take true facts and draw incorrect inferences from them. A trivial example taken from an old (and stale) joke: from the fact that one in every three children born on earth is Chinese, a pregnant, occidental woman with two children anticipates her third child will be Chinese. This error is transparent, but there are many others that are not.

It is generally accepted that "a smoker is more likely to die of lung cancer than a nonsmoker." But since we all die eventually we can make an equivalent statement that has precisely the same meaning but which appears to have very different implications: "diseases *other than* lung cancer are more likely to prove lethal to a nonsmoker than they are to a smoker." While the first statement would likely be made by the American Cancer Society to induce you to quit smoking, the second would likely be made by a tobacco company and would be calculated to have the opposite effect. From the observation that the expected life span of patients who have been hospitalized is substantially shorter than the expected life span of those who have not, are you justified in concluding that you should stay away from hospitals? From the observation that the expected life span of American civilians in wartime is shorter than members of the armed forces should you expect an enlistment in the army to increase your life span?

A subtler variation on this theme arose during World War II. When radar was first developed and installed in British submarines, submarine commanders felt they were encountering enemy aircraft much more frequently than they had earlier. There were a number of possible explanations for this. (The increased observations of German aircraft might have been coincidental, for example.) But the main suspicion was that the Germans had radar detectors that actually attracted enemy planes to submarines using radar and resulted in more frequent contacts. An analysis was made comparing the number of planes one would expect to see using only eye contact with the number of planes that would be seen when eye contact was augmented with radar. The analysis indicated that the increase in the number of contacts could be attributed solely to the radar's greater detection ability. After the war it was established that German planes at this time had, in fact, no radar detection devices.

With a little thought you can avoid gross blunders such as deducing that it is safer to drive at 80, rather than 40, miles per hour from the fact that the number of accidents in which motorists were driving 80 miles per hour is less than the number of accidents in which they were driving at half that speed. (The real question is, of course, what *fraction* of the drivers at each of these speeds have accidents.)

But some problems are thornier—determining whether air or automobile travel is safer, for example, assuming all the relevant statistics are available. Do you measure safety by accidents per passenger mile or accidents per traveling hour? You get very different answers using two apparently reasonable measures. The length of a trip will have a considerable effect on air safety statistics since the most dangerous parts of a flight are the takeoffs and landings. If you triple the length of a flight, you certainly do not triple the probability of having an accident. The calculations for determining risk for automobile travel, on the other hand, are quite different. If you triple the length of a trip, you may well more than triple the chance of an accident because of fatigue.

Testing Your Own Statistical IQ
You may want to test your own skill at drawing valid inferences from known data. In the situations described below you may assume the facts are correct.

We leave it to you to decide what to make of them.

1. Your local newspaper reports that 50,000 people left state X last year and they all settled in one of the other 49 states. (Assume no one else is born, has died or moved.) As a result of this population shift the average IQ in *every* state decreased. Is this possible?

2. The suburbs of Berg and Glen are each connected by rail to the central city of Metropolis. Berg claims it is getting inferior rail service. The claim is tested by sending representatives to the Metropolitan station at random times during the day and checking whether the next departing train is destined for Berg or Glen. It turns out that 90 percent of the time the next train is leaving for Glen. Does it necessarily follow that Berg is in fact underserved?

3. M, B, and S are a money market fund, a bond fund and a stock fund, respectively, and each performs annually according to its own independent probability distribution (which remains the same from year to year). Both the S and B funds have a higher average profit than the M fund but when statistics are compiled to determine which of the three funds had the most profit in a given year it turns out that M wins more than half the time. Is this possible?

4. You walk into a bridge club and the very first hand that you are dealt is one containing thirteen spades—an event that occurs about once in 635 billion hands. Should you assume the deck was stacked?

5. A and B are two companies, each composed of many divisions and each with divisions located in the states of Utah and Ohio

and nowhere else. You have been offered the presidency of both companies and you decide to accept the one with the highest percentage of profitable divisions. A has a higher percentage of profitable divisions in both Utah and Ohio than B. Does it follow that A is your right choice?

6. A person who is afraid of flying hears that it is very unlikely that a person will board his plane with a bomb but very much *more* unlikely that two people will board his plane with bombs. Accordingly, he never boards an airplane without carrying a bomb.

Now let's answer the questions just posed.

1. It may seem counterintuitive but it is possible. Suppose the average IQ in state X is 95, the average IQ in each of the other states is 105 and every one who leaves state X has an IQ of 100. If every state (other than X) receives at least one person from state X the average IQ in every state will be reduced.

2. Berg's complaints may be unjustified despite the apparent weight of the evidence. Suppose the trains to both Glen and Berg each leave hourly so that the service to both cities is identical. But suppose also, that all trains to Glen leave on the hour while all trains to Berg leave six minutes after the hour. An observer would see Berg's train leave first only if he arrived within the first six minutes of the hour and this would occur 10 percent of the time.

3. Suppose that S and B make a 30-million dollar profit 75 percent of the time and a 38-million dollar profit 25 percent of the time. Their average profit of 32 million dollars exceeds mutual fund M's constant profit of 31 million dollars but M will have the greatest earnings of the three more than half of the time.

4. Since the probability of getting a thirteen spade hand is so small, it is tempting to rule out the possibility of a coincidence and to conclude that someone must have stacked the deck. But consider, the probability of getting thirteen spades *is exactly the same as getting any other particular thirteen cards.* Are you to conclude the deck was stacked whatever hand you are dealt?

5. You might think that A is necessarily the better choice but it need not be. Suppose the division breakdown is as shown below:

TABLE 8.1				
	Utah		Ohio	
	A	B	A	B
Profitable Divisions	5	4	23	44
Unprofitable Divisions	20	21	2	6
% of Profitable Divisions	20%	16%	92%	88%

In both Utah and Ohio A has a higher percentage of profitable divisions. But if we view the companies as a whole we have

TABLE 8.2		
	A	B
Profitable Divisions	28	48
Unprofitable Divisions	22	27
% of Profitable Divisions	56%	64%

Although B has a smaller percentage of profitable divisions in both Utah and Ohio it has a higher percentage of profitable divisions when the companies are considered as a whole.

6. We will leave this one for you to work out.

Making Inferences in an Uncertain World

So, if your favorite sport is jumping to conclusions, you had best proceed warily. This is true in the real world in general, and in the financial world, in particular. To invest intelligently you must use the information that is available to anticipate the future. Information about stocks—present and past earnings, dividends, etc.—is easily obtained, but applying that information profitably to find a winning stock is another matter.

Picking an Adviser (in the Stock Market or at the Track)

Instead of trying to outsmart the market by seeking a winning stock you might do better selecting an expert—a mutual fund or individual—and have them pick your stock. Then you need only pick the most competent expert, presumably the one with the best track record. Evaluating stock market forecasters seems to just be a matter of common sense. Just look at past predictions and check how often they are successful and measure their accuracy. But "common sense" can be deceptive. If 30 people are in a room many people think it is more likely than not that they will all have different birthdays; in fact, they will all have different birthdays less than 30 percent of the time. For many financial problems there are many ways of reaching plausible, but false, conclusions and we will examine a few of them.

The process of finding a good financial adviser or a good mutual fund seems simple enough. Look at their records for the last ten years and pick the most successful one. Surely a successful record over ten years cannot be attributed to chance.

But imagine that a thousand stock market pundits making annual predictions in each of ten years are rated on their accuracy. And imagine that in every one of those ten years one particular pundit outperforms a majority of his competitors. If everyone were guessing randomly, such a success would be expected less than one time in a thousand. One in a thousand is pretty unlikely so it seems reasonable to conclude that the successful adviser knows something about the market that most of the others do not.

But now suppose you repeatedly toss a fair coin so that the outcome is random with no particular pattern. And suppose there are thousands of observers trying to guess the unguessable on the basis of what they have observed—the patterns of heads and tails that will come up in the future. If you check every observer's predictions and select the most accurate one, that prediction would also seem very impressive. But the criterion we just used— the frequency with which a person guessing randomly would achieve this success—seems much less persuasive. In the real world there are thousands of mutual funds, financial columnists, and various experts predicting the market, and the person making the best prediction is bound to look good, even if he, as well as everyone else, is guessing randomly. (It is not unusual for a mutual fund that performs in the top 5 percent one year to perform in the bottom 5 percent the following year.) You should not be surprised to win a one-in-a-thousand lottery if you play that lottery a thousand times.

So, if you choose an adviser on the basis of his past record and you want to test his accuracy, test him only on his future performance; his earlier success that initially attracted you to him should be disregarded. Picking an adviser on the basis of his past performance and then using this same past performance to judge him is like telling someone to shoot at a wall and then drawing targets around the places where the bullets hit.

Touts at the racetrack use this principle to their advantage. Suppose a tout claims he has inside information. To prove it he offers to tell you the first-, second-, and third-place finishers in an eight-horse race. If his predictions prove accurate would you pay for information about the following races?

Guessing at random, you would pick the first three finishers once in 336 races, so the tout's accurate predictions are impressive. The probability of picking the winner is $1/8$, the probability of picking the second place horse (after you successfully pick the winner) is $1/7$ and the probability of picking the third place correctly (after getting the first two places right) is $1/6$; the probability of getting all three right is $1/[(8)\,(7)\,(6)] = 1/336$. But if, unknown to you, he has approached 335 other people with all the other possible finishing orders his accomplishment is lessened. Initially your interest may be aroused by the tout's preliminary success, but it is only *after* that success that your testing should begin in earnest.

Consider another case of extrapolating the present into the future. Suppose you are looking for a 'miracle' stock—one that doubles in three years—and you know from past history that only one in every thousand stocks is a miracle stock. You also know from past history that your broker is almost prescient when it comes to distinguishing miracle stocks from ordinary ones. When shown a stock his judgment will turn out correct 90 percent of the time: if he says it is a miracle stock it will be one 90 percent of the time and if he says it is not a miracle stock it will not be a miracle stock 90 percent of the time as well.

If you choose a stock at random, and your broker tells you that it is a miracle stock, what do you think the probability is that you have actually selected a miracle stock, given your broker's comment and our earlier assumptions?

Despite the accuracy of your broker, the stock you have chosen is still a long shot. To see why, imagine that the same situation arose repeatedly—that you picked a stock at random a million different times and try to anticipate what the outcome would be:

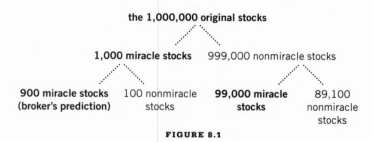

FIGURE 8.1

In Figure 8.1 the original 1,000,000 stocks on the first line are divided into 1,000 miracle stocks and 999,000 non-miracle stocks on the second line. The proportion of miracle to non-miracle stocks reflects the fact that miracle stocks occur once in a thousand times.

Of the 1,000 miracle stocks you would expect 900 to be correctly diagnosed as such and the remaining 100 stocks to be misdiagnosed as non-miracle stocks. Of the 999,000 non-miracle stocks you would expect 899,100 to be correctly diagnosed as non-miracle stocks while 99,900 are mistakenly thought to be miracle stocks.

In all there are 900 + 99,900 = 100,800 stocks that are diagnosed as miracle stocks; notice that there are a great many more non-miracle stocks masquerading in miracle stock's clothing (99,900) than there are real miracle stocks that are recognized as such (900). In fact, the proportion of real miracle stocks (diagnosed as such) to all stocks diagnosed as miracle stocks is $900/(900 + 99,900) = 1/112$.

Your broker's positive recommendation does increase the probability that you have a winner; it changes the probability of success from your original $1/1,000$ to $1/112$ which is almost nine times greater. But considering your broker's predictive accuracy—9 out of 10—a probability of success of $1/112$ seems pretty anemic.

Steven Salop described a similar situation which often arises in a courtroom. In a town in which 85 percent of the taxis are green and 15 percent are blue, a witness testifies that the taxi causing the accident was blue. From past experience it is known that witnesses assign the correct color to a car 80 percent of the time. Despite the 80-percent accurate testimony to the contrary, it is still more likely that the taxi was green than blue. The true probability of being green is $0.17/(.017 + 0.12) = 17/29 = 0.59$.

The First Arc Sine Law

Here is another situation in which intuition may prove unreliable. Suppose you have a friend whom you have been considering as a possible manager of your money for a long time. Your friend has been dabbling in the stock market and has an investment record that goes back many years. At the end of every business day you observe whether or not your friend is a long-term winner or loser as of that date (he is a winner if his overall profits exceed his overall losses *since the first day he traded* and he is a loser otherwise). Suppose you observe that out of a total of 4,000 business days your friend had an overall profit at day's end on all but 100 days, that is, 97.5 percent of the time. How likely is it, if your friend really had no special ability and invested at random, that he could attain such a remarkable record?

The question as stated is hard to answer because the market tends to rise over time. Since the point is to test your intuition we will make a (false) simplifying assumption so that the question can be answered precisely.

Assume that the market is really a single stock that behaves like a random coin toss, its price increasing or decreasing by one, each with probability $1/2$. How likely would it be that 97.5 percent of the time at the end of the business day, the customer's career record would show a profit assuming that he has been observed for a substantial period of time?

According to the first arc sine law[2] if you toss a fair coin many times and, at the end of each toss, check to see if the total number of heads (since you started tossing) exceeds the total number of tails, you would expect that 10 percent of the time the heads would exceed the tails on 97.5 percent or more of the days. And this is so no matter how long you toss.

So it comes to this: in a market with no upward bias, it is much less impressive than it seems to have a high proportion of days when you are a lifetime winner; it is even less impressive when you consider that the market has a positive bias.

Sampling

Inferential statistics is the art of deducing properties about a whole population from a part of it. Although ideally one might want to observe the whole—the entire population—it is usually impractical. If you want to know how much life there is in some old tires in a warehouse you can drive until they are all worn out, but however the experiment turns out you will have no tires left. If you are selling a commercial product it is expensive to poll every potential consumer. What you must often do is extract a random

2. For the mathematically oriented the first arc sine law states that the probability, p, that your friend will be ahead x percent of the time or less over many repetitions is given by the formula

$p = 2 \text{ arc sin } [x/100]^{1/2}/\pi$

sample from a population, examine it, and deduce what you can about the population from which it came. But drawing statistical inferences is a hazardous business.[3]

Why Samples May Be Unrepresentative

What are the hazards of sampling? For one thing people who are polled often lie.[4] They will say they read *Harper's* rather than the tabloid they pick up in the supermarket. Alumni with lower than average incomes may lie or not answer a college's salary survey. Different pollsters get different answers from the same people when questions are rephrased, and in any case the answers they get often depend on the age, race, and gender of the questioner.

If valid inferences are to be drawn from a sample, the sample must be representative of the population which means every member of the population must have the same chance of getting into the sample. In 1936 the *Literary Digest* took a poll to predict the next president. It used the telephone directory and automobile registrations (in the middle of a depression) to obtain its sample—with disastrous results. It predicted Alfred Landon would be the winner, and he barely captured two states.

Sometimes proper sampling techniques seem absurd. Suppose you want to select a random apartment in a building as part of your sample and are initially indifferent to which apartment you choose. *Once chosen*, however, if the apartment turns out to be vacant, you should not substitute a neighbor's apartment (even though the neighbor's apartment would have been acceptable had you chosen it initially). Working mothers, for example, may be heavily represented among the not-at-homes, and they may systematically differ in some way from other apartment residents. Where you take your sample is also important; if you want to know the average weight of the people living in a city, you would not take your sample outside of a Weight Watchers center.

---

3. Even as simple a statement as "The price rose 10 percent one year and dropped 10 percent the next" might induce you to believe the price was ultimately unchanged; in fact it dropped 1 percent.

4. As one wag put it, "The average age of women under 40 is over 40."

Even if your sample is chosen prudently, there is a random element to any sampling procedure and so you might yet be misled. Your best safeguard against distortion by random errors is to take a large sample. But if you are investigating the health of a population and you choose to collect your information near a hospital, you will be in for trouble; and increasing the size of your sample will not save you if your sampling method is fundamentally flawed.

Correlations

When you invest in the stock market you can get a rough idea of what to expect in the future by observing how the stock market behaved in the past. Specifically, if you observe that the stock market rose by 10 percent per year, on average, over the past 40 years, it is reasonable to expect that it will continue to do so. But you also observe that it does not rise 10 percent every year—it rises much more in some years and falls precipitately in others. If your goal is to approximate as closely as possible a 10 percent rise every year, avoiding the large losses in the bad years and foregoing the large gains in the good ones, you might consider diversifying, that is, investing in a large number of stocks rather than in one stock or just a few or all in one industry. By diversifying you hope that the losses in some stocks will be balanced by the gains in others. If you assume that your stocks are what the statistician calls "independent"—that the changes in their prices are uncoordinated—then almost all of the perturbations will be evened out. But the independence assumption means that the behavior of one stock is unrelated to the behavior of all the others and any market observer will testify that stocks tend to dance together. If each stock marched to its own drummer, the market averages would be much more stable than they actually are. If you are concerned about stability, you should be concerned about whether your stocks rise and fall together. A statistical measure of how closely two stocks are related is their correlation.

A statistical measure of how closely two variables are related is their *correlation coefficient*.

The correlation coefficient between two variables is a number between −1 and +1. Suppose you observe two stocks, H and W, for N days. If H_i and

W_i are the prices of stocks H and W on the ith day then the correlation between the prices of H and W is given by the formula

$$\text{Correlation Coefficient} = \frac{1/N[H_1 W_1 + H_2 W_2 + \dots + H_N W_N] - m_h m_w}{s_h s_w}$$

where m_h, m_w are the respective average prices for H and W and s_h, s_w are their respective standard deviations.

If H and W move independently their correlation is zero. If they tend to rise and fall together their correlation coefficient would be positive and if they move in opposite directions their correlation coefficient would be negative. If their correlation coefficient was 1 you could predict the movement of one from the movement of the other—if you graphed the price of one against the price of the other the graph would be a straight line. A correlation coefficient of −1 would also mean the graph is a straight line but as one rose, the other would fall.

A numerically high correlation coefficient—whether plus or minus—means the stocks are strongly related but there is no clue to causality—either stock could cause a change in the other or both might be under the sway of a third factor.[5]

If you want to maximize the stability of a portfolio of stocks it is a good idea to pick not only stocks that you believe will rise but to select either a portfolio of stocks that are almost independent of the market averages or some stocks positively correlated with the market and others negatively correlated with it.

---

5. IQs of identical twins are highly correlated but neither IQ is caused by the other.

Chapter 9

Options

Test Your Intuition 9

1. Stocks A and B each sell for $50 per share now. A call option on stock A permits you to *buy* 100 shares at $50 per share in one year (whatever its actual price at that time) and a similar call option on B permits you to buy 100 shares of stock B at $50 per share in one year. A put option on stocks A and B permits you to *sell* 100 shares of stock A and B, respectively, at $50 per share in a year.

   The price of a call on stock A is higher than the price of a call on stock B. Since calls make money when the stock rises you infer that stock A is more likely to rise than stock B and you infer further, the price of a put on stock A should be lower than the corresponding put on stock B. Any problem with that?

2. A and B are two companies with stocks that are selling for $100. Both A and B are hoping to sign (different) contracts. You know that if A gets its contract its shares will be worth $160 in a year and if B gets its contract it will be worth $140 in a year. If A does not get its contract its shares will be worth $60 in a year and if B does not get its contract it will be worth $10 in a year. Which should have a higher price—a call on stock A or a call on stock B if each call allows you to buy 100 shares of stock at $50 per share? (You may assume that the annual interest rate is 10 percent). See the diagram below.

Price in one year if contract is signed	**160**	**140**
Price today	**100**	**100**
Price in one year if contract not signed	**60**	**10**
	Stock A	**Stock B**

FIGURE 9.1

Answers on page 191.

What Is an Option?

If you read the financial columns you have probably seen the term "deriva-tive." A derivative on a stock is distinct from the stock itself but is related to it; derivatives play the same role in the stock market as side bets in a dice game. Derivatives can serve a wide spectrum of purposes—they can be used to increase leverage (and risks) but they can also serve as an insurance policy that minimizes risks.

The derivatives, or options, that we discuss here will generally be puts and calls on stocks—the right to buy or sell stocks in the future at some fixed price. We will first define puts and calls and show how they might be used either speculatively or defensively. In some special cases we will see how to determine the price of an option and explore some of the relation-ships between the prices of various calls and puts. We will start with an illustrative example, an oil company buying options on some oil fields.

Call Options on Oil Fields—An Example

A field is subdivided into 100 plots, and it is known that on exactly one of them there is $1,000,000 worth of oil. Initially, nobody knows where the oil is, but before six months have passed the oil site will be common knowledge. Since each plot has a different owner, an oil company offers each owner $1,000 for an option to buy his plot. If the plot owner sells this option he receives the $1,000 now and in return the oil company can unilaterally purchase the plot for $500,000 within the next six months, or not, as it chooses. If every plot owner sells an option, each of them will receive a certain $1,000 and in addition, one lucky plot owner will receive an additional $500,000 when his land is actually bought. The oil company will receive oil worth a million dollars in return for paying the plot owners $600,000—a $400,000 profit. Depending upon how they view risk, both the oil company and the owners might consider this arrangement advantageous.

To focus better on the basic ideas, we make our usual assumptions about living in an *Oz*-like world in which there are no commissions or taxes, a world in which you can buy a stock for the same price that you can sell it (or sell it short), and a world where the interest rate is the same whether you make, or take, a loan.

Some Option Terminology

In this chapter we will be concerned with options. The unilateral right to buy or sell something at an agreed fixed price (the *strike price*) at, or perhaps before, some future time, is called an *option*. Options are often associated with stocks and our discussion of options will be restricted to stocks.

A *call* is an option to buy stocks; a *put* is an option to sell them. A typical call option might give you the right to buy 100 shares of IBM stock at $100 per share at any time during the next six months (whatever the actual market price at the time you buy). A corresponding put option might give you the right to sell 100 shares of IBM stock at $100 per share at any time during the next six months (whatever the actual market price at the time you sell).

The price of an option is sometimes called the premium. The price of options are determined by the buyer and seller in the marketplace in the same way that the price of stocks are. As additional compensation, an employer may offer free options to employees but when he does he often restricts their exercise during an initial time period.

The *expiration date*[1] of an option is the last date it can be exercised. The *underlying security* is the stock the option buyer or seller chooses to buy or sell.

---

1. Some options—American options—may be exercised at *any* time prior to the expiration date; European options may only be exercised on the expiration date.

In our oil field example the price, or premium, was $1,000, the expiration date was six months after the option agreement was reached, the strike price was $500,000, and the plot of land corresponded to the underlying security.

Why Buy an Option?

Why would you buy an option rather than buy the underlying security directly? One reason is to minimize risk. Buying an option on a plot of land rather than buying the land itself is a much smaller commitment and in that sense, a much smaller risk. In our example the oil company only paid $1,000 to obtain an option on a plot. It is not clear what a plot would sell for if bought outright, but the expected value of the oil—1 percent of $1,000,000 = $10,000—would probably be the minimum and this is considerably more than the price of the option.

If you believe a stock will decrease in value, you can buy a put or you can sell the stock short. With the latter, the difference in risk is even more dramatic. If you buy a put now (expecting to buy the stock in the future at a low price and resell it at a higher strike price) you can only lose the premium if you guess wrong. If you sell short (selling stock you don't own now but hoping to buy in the future at a lower price) your potential losses are infinite.

A second reason for using options is leverage. Suppose a stock costs $50 and a call option on that stock with a strike price of $50, costs $600. (Such a call entitles you to buy 100 shares of stock at $50 per share.) If you buy 120 shares of the stock at $50 per share for $6,000 and the price of the stock rises to $60, you can sell the shares for a total of $7,200 and you will have made a 20-percent profit. If, for the same $6,000, you buy ten calls and the stock price rises to $60, you can buy 1,000 shares at $50, sell them immediately at $60 per share—a difference of $10,000. Subtracting the $6,000 premium you paid for the options that still leaves you with a net profit of $4,000, or $66^2/3$ percent. If the stock doubles in price, the contrast between stock and option is even better—a 100 percent profit vs. a 733 percent profit. In the real world, commissions and taxes reduce your profit, but clearly when things turn out well you do better buying the options than you would buying the underlying security.[2] Figure 9.2 shows the relationship between the stock price at expiration and the profit/loss made by the purchaser of the call that was just described.

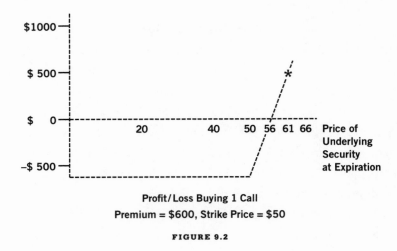

Profit/Loss Buying 1 Call
Premium = $600, Strike Price = $50

FIGURE 9.2

A call may also be used to increase the return on a stock you already own. If you sell a call on a stock you already own (this is called a *covered call*) and if the stock price does not increase before expiration, you are richer by the premium you received for the call.

The Down Side

When you consider that the leverage of calls may increase your profits, that calls are less risky than stocks (at least in one sense) and they may increase the return on stocks you already own you might wonder if there is a down-side to calls, as well. You bet! Each advantage of calls that we mentioned has

---

2. In the real world you would not have to exercise the options and then sell the underlying security. The change in the underlying security's price would be reflected in the option's price and you could sell the option directly on the open market.

TABLE 9.1					
Final Stock Price	$ 0	$ 25	$ 50	$ 75	$ 100
Profit/Loss on Option Investment	–$6,000	–$6,000	–$6,000	$19,000	$44,000
Profit/Loss on Stock Investment	–$6,000	–$3,000	$ 0	$ 3,000	$ 6,000

A Purchase of 10 Call Options at $600 per Option
vs. the Purchase of 120 Shares of Stock at $50 per Share

its down side. If you buy a stock for $50 and that is the price six months later, you may have lost some potential interest income but you have not actually lost money. If you buy a call on this last stock with a strike price of $50 and it expires, you lose 100 percent of your investment. The leverage that gives you the opportunity to make more money for every dollar you risk also raises the probability that you will lose everything (a much less likely prospect when you buy the stock outright).

Finally, if you own a stock and sell a call to increase your profit you lose almost all of your potential profit if the stock subsequently triples in price. In short, buying and selling options, like buying and selling stocks, requires prudence.

What is the Value of an Option?

Having defined what options are, it is reasonable to ask what they are worth. Option pricing is not an easy business. To evaluate an option you must know (among other things) the underlying stock's volatility and the prevailing interest rate (both of which are constantly changing). But the value of an option to *you* is determined not just by the values of certain technical parameters; it also depends on where you sit. A candy manufacturer who buys sugar every year and wants to benefit from any decrease in sugar prices, but also wants to cap his costs if sugar prices rise, might purchase a call on sugar. By paying the premium on a call, he can get the better of today's and tomorrow's prices. And the farmer from whom the sugar is bought can put a floor on the price of sugar by buying puts on sugar; any losses incurred from falling sugar prices would then be offset by the increased value of the puts. And a speculator may accommodate both farmer and manufacturer by selling both a put and call at a price that compensates him for his risk.

Suppose, for example, the current price for sugar is 50 cents a pound and a farmer is fearful that the price will drop in three months when he harvests his crop of sugar beets. If he buys an option to sell his crop at 50 cents a pound (whatever the price at harvest) and pays 3 cents a pound for the privilege, his net proceeds will be $(50 - 3) = 47$ cents a pound if the market price drops below 50 cents a pound at harvest. If the price is more than 50 cents per pound at harvest he will still benefit from the increase. (The farmer and manufacturer may also make a forward contract—that is, set a selling price in advance—but then neither gains if the market turns favorable to himself.) The point is that two parties trying to avoid risk may use different options and a third party (someone who is in a strong financial position who guarantees both sides of the transaction) may deliberately court that risk if he is sufficiently compensated, and all three may be satisfied.

The Factors That Effect Option Prices

Putting subjective judgment aside it is possible to set the price of an option quantitatively by making some reasonable assumptions but the process is too complicated to be described in detail here. What we will do is qualitatively examine some of the elements that determine an option's value.

The value of an option depends upon many factors: the price and volatility of the underlying stock, the strike price, the prevailing interest rate, the expiration date of the option, and the amount and times of the dividends (if any). The precise relationship between these factors and the stock price is far from obvious.

The volatility of a stock is a measure of how much a stock's price changes over time; the greater the volatility, the greater the value of *both* a put and a call. Assume that both stocks A and B cost $100. Suppose also that on any given day stock A rises 10 percent, rises 5 percent, is unchanged, falls by 5 percent, or falls by 10 percent, each occurring a fifth of the time. On any given day stock B rises 2 percent, rises 1 percent, is unchanged, falls by 1 percent, or falls by 2 percent, also each a fifth of the time. The volatility of stock A is greater than that of stock B and the expected value of either a call or a put, respectively, on stock B is less than that of a call or a put on stock A, assuming all strike prices are $100. The point is that the large profit you get when you buy a call and the stock rises precipitately is not balanced by a corresponding loss when the stock plummets. If the stock price at expiration is not higher than the strike price, it doesn't matter to the owner of a call whether the two prices are equal or the stock price is much lower—in either case his call is worthless. In general, the longer the time to expiration of an option, the greater the value of both put and call (assuming there are no dividends).

If all the other relevant factors remain unchanged and you either (i) increase the initial stock price or (ii) lower the strike price—the value of a call will rise, and the value of a put will fall, for obvious reasons. When you pay a dividend you are taking money away from the company and giving it to its shareholders. Dividend payments therefore lower the value of a stock, thus increasing the value of a put and decreasing the value of a call.

As interest rates rise, so does the value of a call. One way to see this is to consider calls and stocks as competing investment vehicles for those who are optimistic about the underlying security. Buying a stock involves tying up much more money than buying a call. As interest rates rise calls become relatively more attractive, and therefore more expensive.

Relationships Between Option Prices

Deriving a formula for determining the value of an option is hard work and we mentioned above that we would not attempt it. But there are certain relationships between different options that can easily be derived from the old adage "there is no free lunch." You may always assume that there is no strategy for buying some options and selling others that will guarantee a profit.

This is not to say that in a real market there are never glitches in which a person can take a risk-free position; such positions may arise but they are anomalies and very rare. But if someone asserts that he has a risk-free, profit-making scheme that he uses repeatedly and that can *never* lose money it is analogous to creating perpetual motion. So if options are priced in such a way that you can make a riskless profit you can be sure those prices are inappropriate.

To see how the "no free lunch" adage is applied, suppose there are three call options on a stock that are identical except for their strike prices which are 35, 40 and 45 and for which we designate the respective call prices as C_{35}, C_{40} and C_{45}. It must then be true that $2C_{40} < C_{35} + C_{45}$. Suppose the inequality were reversed; you could then sell two 40 calls and buy a 35 and a 45 call for a free lunch. Initially you would sell your two 40 calls for more than what the 35 and 45 calls would cost—so you would receive more than you paid. Then, whatever the stock price at expiration, you must either take in some additional money or, at worst, break even. The graph shown below shows how the money you would receive at expiration is related to the price of the underlying stock:

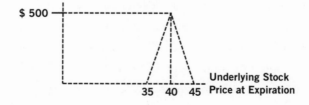

Additional Profit at Expiration

FIGURE 9.3

If the stock price at expiration is below 35, all the calls are worthless. If the stock price is between 35 and 40, you make a profit between 0 and $500; as the expiration price goes from 35 to 40 you get $100 for each $1 increase in the expiration price. From 40 to 45 you lose $100 for every $1 the expiration price rises until you reach breakeven at 45. If the final price is more than 45, no money changes hands at expiration since the amount that you lose on the calls that are sold is balanced by the profits on the calls that you bought.

The Put-Call Connection—A Formula

Common sense suggests that there should be a relationship between the price of a call and the price of a put, and in fact there is one; but, as we indicated earlier, that relationship is just the reverse of what you might guess. Imagine that calls on a stock are unusually expensive. Since you make money on calls when the underlying stock rises, a high call price would seem to indicate that the market senses the stock is about to rise. And since you make money on puts when the underlying stock falls, this seems to further imply that puts should be cheap. But in fact the prices of puts and calls do not move in opposite directions, they rise and fall together. Let's take a closer look to see why.

Suppose you buy 100 shares of a stock that costs $50 and a put on that same stock with a strike price of $50. Suppose also that there is a 6-percent interest rate on money during the life of the put. It is easy to verify that you have in effect bought a call (we ignore commissions and taxes). If the stock is worth, say $80, at expiration and you sell it, your profit on the stock is $100 \times (\$80 - \$50) = \$3,000$ and you do not exercise your put; if the stock is worth $30 at expiration, your loss of $2,000 on the stock is exactly balanced by what you receive when you exercise the put. This is precisely what happens when you buy a call: you gain $100 for each dollar that the stock gains and are at breakeven when the stock fails to increase.

What is the cost of taking this position (evaluated when the options mature)? You must buy the put initially at a price P and pay interest on it for the life of the contract a total of $P + 0.06P$. You must also pay the interest on the cost of the stock, 0.06 (100 × $50). (The cost of the stock itself is received when you sell it.)

If you bought the call directly at a price C your cost would be the price of the call and the interest on it which would be: $C + 0.06C$.

Since the two position are identical they must have the same price; if they didn't you could buy one position, sell the other, and make a sure profit. It follows, therefore, that $C + 0.06C = P + 0.06P + 0.06 \ (100 \times \$50)$ or $C = P + 0.06 \ (100 \times \$50) / 1.06$.

In general, suppose you buy 100 shares of stock at S per share and also buy a put on that same stock with strike price S. When the put is about to expire, you exercise the put if appropriate and sell the stock. If you ignore the commissions and taxes, and observe what happens for various stock prices at expiration, you will see that you have in effect bought a call with strike price S. For each \$1 the stock rises above S you make \$100 on the stock (the put is worthless); if the stock falls you are at breakeven, your loss in the stock exactly balanced by your gain in the put.

If R is the interest rate for the period of the option, how much does it cost to take this position (evaluated at the put's expiration date)? First you must buy the put for P and hold it until expiration—a total cost of $P(1 + R)$ where RP is the cost of interest for the duration of the put. You must also pay the interest on 100 shares of stock, that is, $100RS$. The total cost of taking this position is $P(1 + R) + 100RS$.

If you simply buy a call, your total cost is the price of the call plus one year's interest on it or $C(1 + R)$. Since both positions are equivalent we conclude that the costs should be the same, that is, $P \ (1 + R) + 100RS = C(1 + R)$ or $C = P + (100RS) / (1 + R)$. Clearly, as the put price increases (or decreases), so does the call price. Notice also, that if the interest rate is 0 percent the values of the call and the put are the same, that is if $R = 0$, $C = P$. Observe that if you write the equation in the form $C = P + (100S) / (1/R + 1)$, as R increases, the denominator on the right gets smaller so the gap between P and C increases.

What if the actual market price of a call is higher than the price indicated by the formula? You could then sell the call and buy both a put and 100 shares of stock at the market for an immediate profit and a riskless position. At expiration, your holdings would be completely neutral and so a profit would be assured. If the price of the call is lower than that prescribed by the formula, you could buy the call, sell the put and sell short 100 shares and once again, your profit would be assured. Again, we stress that we are talking about a world with no commissions and taxes, and a world where one can buy or sell at the same price and borrow or lend at the same interest rate (of course anyone that lives in the real world knows the marketplace is not so generous).

The Apparent Put-Call Paradox

What was wrong with the logic of our earlier argument that concluded that expensive calls implied cheap puts? The key factor in option pricing is *not* buyer optimism or pessimism—that is already reflected in the price of the underlying security. A high option price for an option—whether it is a put or a call—reflects volatility and not optimism/pessimism about the stock's prospects.

There is also a relationship between the prices of put and call options that have the same strike price when that strike price differs from the underlying stock price, but it is a bit more complicated. Before we examine the general case we will look at a particular numerical example.

Suppose you buy a stock that costs $90, you buy a put on that stock that has a strike price of $84, and, in addition, you borrow $8,000. Suppose also, the interest for the period of the call is 5 percent.

Then suppose the stock at expiration is worth $95. Your put would be worthless, your stock would be worth $9,500, and you would have to repay your loan with 5 percent interest, that is, $8,400. Your net worth would then be $9,500 − $8,400 = $1,100.

Now suppose the stock at expiration is worth $75. The put would be worth ($84 − $75) (100) = $900, your stock would be worth $7,500 and you would have to repay a debt of $8,400.

Notice that the effect in both of these cases would be exactly the same as if you had bought a call with a strike price of $84.

Now assume that C, P are the prices of a call and put, respectively, with a strike price of \$84. What is the cost of taking the *initial* position? You would have to pay for the put (P) and the difference between the stock you bought (\$9,000) and the amount of your loan (\$8,000). The initial cost of a call we said was C so $C = P + \$9,000 - \$8,000 = P + \$1,000$.

Now consider the general case. Suppose C and P are the prices of a call and put, respectively, on the same stock with the same expiration date and with the same strike price of K, where K may differ from the stock price S. Assume the stock price at expiration is S^*, and R is the interest rate for the option period. Imagine that you

    i. buy a put,
    ii. buy 100 shares of stock, and
    iii. borrow $100K / (1 + R)$.

At expiration you sell the stock at S^*, exercise the put if appropriate, and repay the loan.

If $S^* > K$, the put is worthless—you will have $100S^* - 100K$: this is the value of the stock less the loan (with interest) that you have to repay.

If $S^* < K$, you will have nothing; the put will be worth $100K - 100S^*$, and when added to the value of your stock ($100S^*$) you will have $100K$: just enough to repay the loan.

In effect, your position is equivalent to a call with strike K. Its initial cost is $P + 100S - 100K / (1 + R)$ so that the formula becomes $C = P + 100S - 100K / (1 + R)$. Notice that if $K = S$ this formula reduces to the one we derived earlier.

Setting Futures Prices

Suppose that a buyer and seller wish to make a sale of 100 shares of stock at some future date, but set the price now. One can use puts and calls to determine what that future price should be (assuming the expected value of both buyer and seller will be zero).

Given a put and call on a stock with the same strike price K and expiration date, it is clear that for a sufficiently large K the put will be worth more than the call, and if K is zero the call will be worth more than the put. For some intermediate value of the strike price K, the put and call will have the same price. At that price both the buyer and seller of the futures contract have an expected profit of zero. Substituting $P = C$ in the formula $C = P + 100S - 100K / (1 + R)$, we obtain $K = S(1 + R)$ which is not very surprising; if you make an agreement to buy or sell a stock in the future, the future price should be today's price plus interest.

As we said earlier, the derivation of option prices is complex in general but in special cases, when you are given certain information, the derivation can be straightforward.

Setting Put, Call Prices with Greater Information

Suppose a company's stock is worth $60 today. The company has just bid on a contract and if it is successful you know (somehow) the price of the stock will rise to $100 in one year; if the bid is unsuccessful, the price of the stock will fall to $40 in a year. The prevailing interest rate is 10 percent.

Can you derive the appropriate price of a put, P, and a call, C, each with a strike price of $50 solely on the basis of these assumptions? (Notice that you do *not* know the probability that the contract will be signed.)

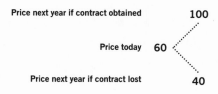

Price next year if contract obtained   **100**

Price today   **60**

Price next year if contract lost   **40**

FIGURE 9.4

At first it might seem that such a calculation is impossible. The probability of winning the contract seems to be a critical factor in determining the value of a put or a call and that information is not available. Your own subjective assessment of the option values would also be influenced by your personal attitude toward risk, about which we have said nothing. But, remarkably, you can derive the appropriate price of an option without being told the relevant probabilities (in a sense the probabilities are implicit in today's stock price) and those derived option prices will be compelling *whatever* your attitude toward risk. If the price of the put or call deviates from the prescribed price you can take a hedge position which will guarantee that you will never take a loss and will likely get a profit: a free lunch.

The first step in the derivation is to set up an arbitrage position: you borrow B, sell one call and buy N shares in such a way that

a. you borrow exactly enough money to buy the N shares after deducting the money that you receive for the call;

b. you break even if the contract is not signed;

c. you break even if the contract is signed.

If you translate these verbal conditions into mathematical equations you obtain respectively,

a. $B = 60N - C$

b. $40N = 1.1B$; the value of the stock, 40, is less than the strike price, so the call is worthless. Therefore the value of your stock must be enough to pay off your loan with interest.

c. $100N - 100 (100 - 50) = 1.1B$; the value of the stock less the value of the call you sold must be enough to repay your loan.

Using elementary algebra, you can solve for N, B and C. It turns out that $B = 200{,}000/66 = 3{,}030.3$, $N = 500/6 = 83.3$ and $C = 130{,}000/66 =$

1,969.7—that is, you should borrow about \$3,000, you should buy about 83 shares, and the value of the call is just under \$2,000.

Suppose the market is not clever enough to work all this out and calls actually sell for more than they should, say \$2,500. By following exactly the same procedure we just described you would have exactly the same outcome except that when you are done you will be \$2,500 − \$1,970 = \$530 richer. If calls sell for less than they should, say \$1,500, you reverse the procedure—you buy the call, lend \$3,030 (assuming you can lend the money at the same 10-percent rate at which you borrowed it) and sell 83.3 shares. This guarantees a profit of \$1,970 − \$1,500 = \$470.

So if your assumptions about the stock's price next year are correct, you can act so as to make a certain profit unless the puts and calls are correctly priced, and this is true without knowing how likely it is that the contract will materialize. Your attitude toward risk is also irrelevant—all that is necessary is that you prefer to make a profit rather than take a loss.

You can derive the value of a put with a strike price of 50 in a similar way. You buy N shares of stock, you also buy a put and you borrow enough money, B, to finance both purchases. The equations to be satisfied are:

a. $B = 60N + P$
b. $100N = 1.1B$
c. $40N + 100 (50 - 40) = 1.1B$

Equation (a) states that you must borrow enough to pay for the stock and the put. Equation (b) states that the value of your shares must be enough to pay off your loan if you get the contract since your put is worthless. And equation (c) states that the combined value of you stock and put must be enough to pay off your loan if you do not get the contract.

The solution is $B = 100,000/66 = 1,515.2$, $N = 100/6 = 16.7$ and $P = 34,000/66 = 515.2$—that is, you should borrow \$1,515, buy about 17 shares and the put's value should be \$515.

Option price calculations may seem counterintuitive and lead to apparent paradoxes. For example, suppose there are two one-year call options on two different stocks and each has a strike price of 50. The interest rate is at 10 percent and the present price of each stock is 100. At expiration, stock A will either be at 60 or 160 and stock B will either be at 10 or 140 (in both cases the price will be determined by whether the company is awarded a contract). Would you pay more for a call on stock A or a call on stock B? You are not told the probability of either contract being signed and you know nothing about the purchaser's attitude toward risk.

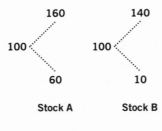

FIGURE 9.1

Since both the upper and lower prices are higher for stock A you might think this would be the pricier call option but it is not. The price of a call for stock A should be $5,000. If that is the price of the call and you borrow $6,000 and sell a call you will have enough to buy 110 shares of stock as well. At expiration, if the price of the stock is $160, you can sell your 110 shares of stock for $17,600, buy the call for $100 (160 − 50) = $11,000 and still have just enough to pay off your loan with interest: $6,600. If the stock price at expiration is $60, the call is worthless and you will have $6,600—just enough to pay off your loan with interest. (If the market sets the call price lower than $5,000, you have a riskless hedge— you sell short 110 shares of stock, buy the call, and lend the surplus for a sure profit; if the call price is too high, do the reverse: sell the call, buy 110 shares of stock and borrow $6,000, to the same effect.) Using similar logic, you can show the call on stock B is worth $6,293.70. In that case, you buy 69.23 shares of stock, and borrow $629 (approximately). If the stock is

$140 at expiration the call is worth $9,000, the stock is worth $9,692, and this leaves you about $692 to pay off your loan. If the stock is worth $10 at expiration, the call is worth nothing and your stock, worth $692, is just enough to pay off your loan.

An intuitive explanation why a call on stock B is worth more than one on stock A is this: since both stocks have the same market price and stock A's final prices are higher, the market must consider it more likely that B will end at its higher price than that A will. But whether you accept this argument or not, the no-risk arbitrage argument is unassailable.

With two possible future prices and today's stock price you can generalize the formula. Suppose

S = the price of a stock today

H = the higher of two possible prices of the stock in a year

L = the lower of two possible prices of the stock in a year

K = the strike price of both the call and the put

R = the prevailing interest rate for a year

P = the proper price of a one-year put

C = the proper price of a one-year call

We assume $L < K < H$.

For the call, you make the same hedge you made earlier: you borrow B, sell a call and buy N shares to create a neutral hedge. Again you must satisfy $B = NS - C$; $(1 + R) B = NL$; $(1 + R) B = NH - 100 (H - K)$. These equations state that you borrowed enough and will break even whether you get the contract or not. The solution turns out to be

$$N = 100 \frac{H-K}{H-L}; \quad C = 100 \frac{H-K}{H-L} (S - \frac{L}{1+R}); \quad B = 100L \frac{L}{(H-L) + (1+R)}$$

To derive the put price you create a neutral hedge by borrowing B, buying a put, and also buying N shares. The equations that must be satisfied are $B = NS + P$; $(1 + R)B = NH$; $(1 + R)B = NL + 100(K - L)$. The solutions are

$$N = 100\ \frac{K-L}{H-L}\ ;\quad P = 100\ \frac{K-L}{H-L}\ (\frac{H}{1+R} - S);\quad B = 100H\ \frac{K-L}{(H-L)+(1+R)}$$

If you calculate $C - P$ and simplify, you obtain $C = P + 100S - 100K / (1 + R)$, the formula we derived earlier.

There are a few points about our derivation of the option prices that are worth repeating. The critical issue in the option-pricing problem seemed to involve the probability of attaining the contract and the value of the options seemed to depend upon a subjective attitude toward risk. It turned out, however, that the option prices were derived without mentioning the probabilities at all. Risk aversion turned out to be irrelevant because you could take a hedge position that would guarantee a sure profit if the prices were set erroneously. The probability of getting the contract was implicit in the price of the stock today since you can view today's price as a future prediction. But the remarkable thing is that you do not really care whether the market is a good predictor—you can take an arbitrage position based on this market "prediction" whether it is accurate or not. You do have to be right about the consequences of receiving, and not receiving the contract, however; that is, the stock price must go to 100 if you get the contract and drop to 40 if you do not. But if your predictions are correct about these two future prices this arbitrage position doesn't just put the odds in your favor, it allows you to hedge in such a way that you cannot lose.

Appendix

Test Your Intuition: Answers

CHAPTER 1

1a. Suppose you have X dollars and always bet 75 percent of it. Half the time you will win and have $2.5X$ after the bet, and half the time you will lose and have $0.25X$. On average, you will have $1.375X$. After 5,000 bets you will have $\$100(1.375)^{5,000}$, on average, which in dollars is more than a 1 followed by 690 zeros.

1b. When you have X dollars and win the next bet you have $2.5X$ after the bet (you bet $0.75X$, you win $1.5X$ and together with your original X you have a total of $2.5X$); when you lose, you have $0.25X$. If you win n of your 5,000 bets you will have, $\$100(2.5)^n(0.25)^{5,000-n}$. If you are to break even after 5,000 bets, $(2.5)^n(0.25)^{5,000-n}$ must be at least 1 and n must be greater than 3,010. The probability of winning more than 60 percent of the time when you make 5,000 even bets is extremely small. It is less than $1/k$ where k is a 1 followed by 44 zeros.

1c. Since you win half the time, $p = 0.5$ and since you win twice the amount you risk, $W = 2$. The optimal bet is therefore $f = (pW - 1 + p)/W = 1/4$; you should always bet a quarter of what you have.

2a. Your chance of leaving a winner is 0.00004.

2b. Your probability of winning would become 0.268 if you made $10 rather than $1 bets.

3a. Even when your chance of winning each bet is as high as 0.49 if you start with $50 and make $1 bets you will leave a winner less than 12 percent of the time.

3b. If you start with $87 you are more likely to leave a winner than loser but just barely—the probability of leaving a winner is 0.587.

CHAPTER 2

1. The average annual rate of increase in value between 1626 and 1986 is about 8 percent. To confirm this, observe that $\$24 \times (1.08)^{360} = \$25,868,000,000,000$.

2. One percent interest per day is very different from 365 percent interest per year; in fact it is more than 10 times as much. Your lump sum payment should be $1,000 \times (1.01)^{365} = \$37,783$ in a non leap-year. This is equivalent to a real annual interest rate of about 3,778 percent.

3. In the long run you will make about 14.87 percent on your investment (not $^1/_{10}$ of 300 percent = 30 percent).

CHAPTER 3

1. If, at any given time, the interest rate was the same on all bonds whatever their duration, it would be possible to buy and sell bonds in such a way that a profit would virtually be inevitable and a loss, impossible. You can review the details in Chapter 3 if you like.

2. If the interest rate changes just once as described in the text, you will have more money than you anticipated at the time of purchase however interest rates might change. The only time you will not be ahead will be if interest rates remain unchanged.

3. If the bond paid no dividends rather than a $50 dividend, the value of the bond would decrease by 24.4 percent.

CHAPTER 4

1a. You will repay about $600,000. Of this, $100,000 will be repayment of the original loan and the remaining $500,000 will be interest.

1b. If you take a 36-year, rather than a 40-year, mortgage you cut the period of repayment by 10 percent. This will increase your mortgage payments, of course, but only by $2/month or about a sixth of 1 percent.

1c. After repaying for 36 years (90 percent of your original loan) you still owe a considerable portion of that loan: about $45,000 or 45 percent of it.

1d. Part of any payment of your loan will be interest on what you still owe and the rest will lower your debt. Your first monthly payment will consist of 99.7 percent interest and only 0.3 percent principal.

2. By increasing the amount you repay your debt by 3 percent every year you will lower your first monthly payment to $520: a decrease of more than 27 percent.

CHAPTER 5

1. Clearly, the lower the inflation rate, the longer your pension will last. If the inflation rate is 5 percent, you will be covered for about 22 years before your money runs out. If inflation drops to 4 percent, your money will last about 25 years. If inflation drops to 3 percent, your money will last about 29 years. If inflation drops to 2 percent, your money will last about 37 years.

2. By saving $10,000/year for 45 years you will accumulate almost $800,000 ($791,000).

3. If you are to live on your pension forever, you may withdraw 55 percent of your pension ($27,500) or less during your first year of retirement.

CHAPTER 6

There really is no short answer to this question. Its real purpose is to show the analogy between the prospective investor in a wild bull market and the prospective buyer of the magic bottle. The reasoning in both cases is similar. The purchaser of the magic bottle feels he can hardly lose if he sells the bottle soon after he buys it and acquires whatever worldly goods he desires in between. The stock buyer makes his profit rapidly, getting in and out before time has a chance to tame the raging momentum of the bulls. The bottle buyer insists on buying at a high price so that he'll have no trouble convincing someone else to buy a bit lower; the stock buyer, though he recognizes there is a manic, unsustainable quality to a market rally, is confident that he is buying in the early stages and will sell long before the sky falls. And in both cases it is clear that someone must get hurt. In the case of the bottle any one selling price can only be succeeded by a finite number of subsequent selling prices and then the axe must fall; in the stock market the mathematical certainty is missing but inevitably the shooting stars must return to earth. And in both cases someone who was confident that he could not lose will finish a loser.

CHAPTER 7

1. Both parties are advised to choose their strategy by using a random device rather than simply picking one of their obvious options; this will maximize their average return. The detective agency's strategy seems reasonable: guard the bank with the most money almost all of the time (with probability $100/101$) and guard the bank with the least money less than 1 percent of the time (with probability $1/101$). On average they will lose $990,100. Notice this is a bit less than the loss they will have if they always guard bank II and the thieves always rob bank I. Oddly enough, the thieves do best to rob the bank where the money isn't most of the time. They do best to rob bank I with probability $100/101$ and bank II the rest of the time. The outcome will be an average loss of $990,100 if either party adapts the recommended strategy and this will be true whatever their adversary does.

2. No matter how many diamonds there are you can make the probability of choosing it as high as 0.3678 if you go about it the right way. Since you know the number of diamonds initially you can reject the first 36.8 percent of them initially and then choose the first diamond that is larger than all the diamonds that preceded it.

CHAPTER 8

1. Notwithstanding your broker's diagnostic skills the probability the chance that your stock is a miracle stock is still pretty slim (although better than it was before he spoke). To see why, imagine that your broker had diagnosed 10,000 stocks and decide how many of them you would expect to be miracle stocks. Now compare the number of incorrectly diagnosed non-miracle stocks with the number of correctly diagnosed miracle stocks (999 versus 9).

2. Even if a poker player were no better or worse than his associates and even if he played for a very extended time period it would not be unusual for him to be ahead (or behind) a large fraction of the time. There is a great deal of "inertia" that allows those who are ahead/behind to remain ahead/behind. In particular, a player would be expected to be ahead on 97.5 percent of his playing days about 10 percent of the time.

CHAPTER 9

1. The price of calls and puts on the same stock tend to rise and fall together. If news favorable to a corporation is revealed it's stock will tend to rise; but if there is uncertainty concerning a stock—a possible strike, a critical contract to be awarded—which make precipitous future changes likely (up or down), its volatility will increase and this will increase the value of both puts and calls.

2. Although both the possible future high price and the possible future low price of stock A are higher than the corresponding prices for stock B, the value of a call on B is worth more than the value of a call on A. Intuitively, you can say that the equal market price of the stock today (100) is telling you that B is much more likely than A to rise. It turns out that the price of a call on A should be $6,000 and that the price of a call on B should be $6,294. The final justification for these call prices (which is included in this chapter) are not based upon intuition, however. If the market price of a call differs from the prices we have quoted there must be a riskless arbitrage position which will guarantee a profit.

Index